WIND

The Earth series traces the historical significance and cultural history of natural phenomena. Written by experts who are passionate about their subject, titles in the series bring together science, art, literature, mythology, religion and popular culture, exploring and explaining the planet we inhabit in new and exciting ways.

Series editor: Daniel Allen

Wind

Louise M. Pryke

REAKTION BOOKS

To my father, Grahame, who loves to have the wind in his sails

Published by Reaktion Books Ltd
Unit 32, Waterside
44–48 Wharf Road
London N1 7UX, UK
www.reaktionbooks.co.uk

First published 2023

Copyright © Louise M. Pryke 2023

Printed and bound in India by Replika Press Pvt. Ltd

A catalogue record for this book is available from the British Library

ISBN 978 1 78914 720 9

CONTENTS

Introduction

Wind is a subtle natural phenomenon, with an elusiveness only matched by its tremendous power. Unlike many other climatic phenomena, such as rain, fire or storms, wind cannot be observed by the naked eye. Instead, it must be experienced through other senses such as touch and hearing, or 'seen' through the lens of its effects on our surroundings. Wind, the movement of air from areas of high to low pressure, maintains a perpetual motion by definition: it is as ancient as it is changeable, as omnipresent as it is intangible.

From Earth's earliest origins, wind has shaped our planet – shifting the foundations of the physical environment and setting the course of human civilizations. Wind has played a vital role in the development of Earth's environment and atmosphere: recent studies have shown the powerful effects of wind-enhanced tectonics even in the movements of Earth's plates.

In human history, the shifting patterns of wind have influenced evolution and culture. Perhaps for this reason, wind is conceptually fused with change in human thought, with shifting fates connected to the movement of air in innumerable works of poetry, music, literature and myth. The role of wind in human history is itself in a state of change in the modern day. The ability of wind to be farmed for the provision of clean power and to combat climate change has become a dominating, at times divisive, topic.

By turns creative and destructive, wind spreads seeds, fills sails and disperses the energy of the Sun, allowing for a habitable

Utagawa Hiroshige, *The City Flourishing, Tanabata Festival*, no. 73, from the series 'One Hundred Famous Views of Edo', 1857, colour woodblock print.

7

Thomas Birch, *Loss of the Schooner 'John S. Spence' of Norfolk, Virginia (2d View – Rescue of the Survivors)*, 1833, oil on canvas.

Claude Monet, *Rising Tide at Pourville*, 1882, oil on canvas.

Spring storms light up
a wind farm in Indiana.

biosphere. Its formative effect on our planet's environment is reflected in the phenomenon's prominent role in myths and religions of antiquity. Wind's ubiquity across diverse cultural and historical environments has resulted in manifold human responses. A common theme in cultural and religious responses to wind has been a recognition of the potential for wind to support and spread life on Earth and, at times, risk its destruction. Wind has inspired groundbreaking scientific innovations and has appeared in works as diverse as the Hebrew Bible, the poetry of Keats and the blockbuster film *Twister*.

In the twenty-first century, the necessity of understanding and appreciating the natural world has become increasingly clear. This drive towards further study also invites an exploration of the connection between nature and culture. Wind has been worshipped since antiquity, decided the outcome of innumerable battles and powered the processes of planetary formation, yet it remains intangible and unpredictable. Delving deep into the world of the wind means encountering an ephemeral reflection of nature's complexity.

Wind plays a complex yet critical environmental role, and its symbolic status in human thought is equally nuanced. This book explores the topic of wind and its image throughout history. In the process, we gain a clearer view of the invisible yet irresistible force of wind.

Ventus'midionalis. 2plo. ca. in£. hu. imp. Electo transetes plonu regice: .uuam, ofe:r pecroti. Nocuitum ꝫturbat sensus. Remo nocti cum camfora ꝫ aqua rosea. Conueit. fi:h. sie te crepitis. autumpn oꝛ septentonalibꝫ.

1 Wind: Natural History

Wind has been a constant yet ever-changing presence through-out human history. Defying its ethereal nature, wind is as essential to life as any other natural phenomenon. As well as spreading life through the dispersal of seeds, spores, water vapours and pollens, wind has not only influenced the growth of plants and the evolution of their physical forms, but shaped the behaviours, metabolisms and forms of animals. Despite its subtlety, wind is a powerful terraforming force, one as influential as glaciers and rivers in the moulding of landmasses and mountains.[1]

Along with natural history, this chapter considers the history of human studies of wind. In numerous scholarly works, including Aristotle's *Meteorologica* (340 BCE), the observation and study of wind have attracted scientists, philosophers, astronomers and authors from diverse backgrounds and historical settings. From Francis Bacon's *History of the Winds* (1622 CE), to the surprisingly intricate development of the Beaufort scale by which wind is measured, writing about wind has proved a stimulating companion to the development of modern climate science.

South wind, illumination from *Tacuinum Sanitatis* (1380–99), a medieval health manuscript originally from the 11th century.

Defining wind

Simply expressed, the word 'wind' describes the movement of air from areas of high pressure to areas of low pressure in the atmosphere. The speed of this movement of air is increased where the differences of pressure are greater. Air flows clockwise around a high pressure system, and counterclockwise around a low pressure system.

Wind is a force largely invisible to the naked eye. It exists in numerous forms, and its behaviour is frequently unpredictable. It is natural, then, for attempts to define wind to involve some ambiguity. The science of studying wind is a rapidly developing field, with many exciting new discoveries in recent years – yet much about this ethereal, fluid force remains up in the air.

The energy for the movement of wind around the atmosphere comes from the Sun. While it is generally accepted that wind is solar powered, understanding the generation of wind in the environment and its place in atmospheric circulation is a complex and constantly developing area of study.

Wind plays an important role in Earth's water cycle. In recent years, the once-accepted view of the water cycle has changed, and wind's role in atmospheric circulation has found greater appreciation. Traditionally, it was thought that the planet's water cycle mostly involved evaporated moisture from large bodies of water being distributed by wind across the Earth's surface. In the last fifty years, that view has changed – it is now known that forests play a significant role in generating the transpiration that drives rainfall. The various plants that live within forests capture water in their roots, and then release it as vapour through transpiration. Prevailing winds transport the water through the air, delivering rainfall to other locations. When this water moves through the power of wind in large volumes, it becomes what meteorologist José Marengo called a 'flying river'.[2]

One of the world's best-known flying rivers is found in the Amazon basin. Trees in the Amazon rainforest draw water from the ground and pump massive quantities of water into the atmosphere. The better-known terrestrial Amazon River that flows to the Atlantic carries 17 billion tonnes of water a day, making it the world's largest river by volume (though shorter in length than the Nile). However, the invisible flying river that flows above the Amazon canopy carries an even greater volume: an estimated 20 billion tonnes of water a day. Flying rivers are found everywhere, with a trans-Siberian flying river delivering

The Amazon rainforest.

around 80 per cent of China's rainfall. The massive movements of water involved in flying rivers, in the Amazon and beyond, are shaped into streams transported by the wind, bringing rain.

Recent research has posited that as well as carrying flying rivers, wind may also be created by them. It is thought that forests may act as a 'biotic pump' by generating winds through changes in air pressure relating to the process of transpiration.[3] The 'pump' is fuelled by water vapour formed through condensation over coastal forests, creating areas of low air pressure, which generate winds. Continuing cycles of transpiration and condensation can then produce winds that deliver rains thousands

of kilometres inland.[4] With at least 40 per cent of the world's rainfall coming from the land rather than the ocean, deforestation in areas that supply the world's flying rivers is a critical issue for environmental conservation. The loss of circulating moisture caused by the destruction of forests has been identified as an imminent threat to the well-being of the planet, comparable to climate change.[5]

Wind in Earth's history

Wind has played an often-overlooked role in human evolution, having influenced human movement from prehistoric times. The causes underlying the emergence of human traits during the evolutionary process remain a topic of debate.[6] Yet the need for early humans to adapt to their environments holds general acceptance, even as the exact mechanisms of this adaptation remain under investigation. A current theory credits wind, among a convergence of influences, with assisting our human ancestors in their transition from moving on all fours to standing upright. Early hominids who evolved an erect stance may have benefited from a reduction in heat stress in warm environments. Air moving around erect hominids would have exposed them to evaporative cooling far more than if they were closer to the ground. Conversely, wind chill may have disadvantaged Neanderthals. Archaeological research suggests that Neanderthals were physiologically cold-adapted, but in comparison with early modern humans, their clothing was less adaptive to rapid changes in climate – significantly, to the effects of wind chill.[7] While it is difficult to ascertain the levels of wind to which the Neanderthals were exposed, the rapid cooling of Earth's poles during ice ages likely corresponded with a combination of cold and wind spikes. This combination would have left the Neanderthals vulnerable to hypothermia, while the clothing of early modern humans was better suited to blocking wind penetration, as archaeological records suggest it was more complex in its manufacturing.[8]

As well as influencing the mechanics of motion, wind has played a role in how and where humans live. The need to find shelter from strong winds, storms and temperature fluctuations was a driving force in the development of shelters for early humans. Similarly, environmental fluctuations, the shape of the terrain and the distribution of plants to provide food would have influenced the patterns of distribution of humans over the globe. Finally, for as long as humans have raised sails, winds have carried humans and their cultures to new horizons.

The history of studying wind

Moved by the powerful presence of wind, numerous literary works have been devoted to studying, measuring and understanding wind, from antiquity to the modern day. The ubiquity of wind sees it feature in many early myths and legends, as well as early scientific observations. Many classical works addressed the topic of anemology (the study of winds), notably the extended explorations of the subject in Aristotle's *Meteorologica*, Theophrastus' *On Winds* and Seneca's *Natural Questions*.

Debates over the nature of wind – still in progress today – were already a lively area of scholarship when Aristotle was writing *Meteorologica* around 340 BCE. Presocratic scholars, such as the earlier Greek philosophers Anaximander, Anaximenes and Heraclitus, considered wind to be part of air, earth and water – all of one essence, but taking different forms. For Anaximenes, wind was part of an elemental explanation of the cosmos, connected to air as well as breath. With breath and wind both being subject to 'blowing', they were considered to be related to exhalations and inhalations. Anaximenes and his predecessor Anaximander saw wind as the cause of lightning, through rupturing clouds.[9]

Aristotle made a firm departure from the theories of the Presocratic philosophers. Rather than conceiving of wind as the movement of air, Aristotle believed that wind was a dry, hot exhalation from the earth:

> There are some who say that wind is simply a moving current of what we call air, while cloud and water are the

same air condensed; they thus assume that water and wind are of the same nature, and define wind as air in motion. And for this reason, some people, wishing to be clever, say that all the winds are one, on the ground that the air which moves is in fact one and the same whole, and only seems to differ, without differing in reality, because of the various places from which the current comes on different occasions: which is like supposing that all rivers are but one river. The unscientific views of ordinary people are preferable to scientific theories of this sort.[10]

Raphael, *School of Athens*, c. 1508–11, wall fresco, Room of the Segnatura, Apostolic Palace, Vatican City.

Aristotle's rejection of the theories of Presocratic philosophers incorrectly conflates their views: while Anaximenes believed that wind, clouds and rain were essentially compressed air, this view was not held by other philosophers.[11] Despite its inaccuracies, Aristotle's view was hugely influential on early studies. For almost 2,000 years – until *History of the Winds* was published by the philosopher and statesman Francis Bacon in 1622 – Aristotle's theory of dry-hot terrestrial exhalations dominated the scholarly study of wind.

Despite the dominance of Aristotle's views in antiquity, other ancient writers continued to explore the topic and contribute to the study of wind, notably Theophrastus, Seneca, Varro, Pliny, Strabo and Vitruvius. Vitruvius, like Seneca and Hippocrates, did not subscribe to Aristotle's views on wind, instead seeing wind as a 'flowing wave of air'.[12] The influence of different types of wind on human health, as considered by Hippocrates and Vitruvius, continued to be debated during the Renaissance, and remains a developing area of research in the present day.

The views on anemology of Theophrastus, Aristotle's successor in the Peripatetic school of Greek philosophy, have often been conflated with those expressed by his famed predecessor. Though clearly influenced by earlier works, Theophrastus' *On Winds* makes several departures from the concepts outlined by Aristotle. Theophrastus' interest in the balance between opposing forces informs his exploration of how wind affects health, behaviour and the natural environment – considerations that continue to capture the interest of climate scientists today.[13] Wind was further explored in the Roman philosopher Seneca's *Natural Questions*. Written around 65 CE, *Questions* considers meteorological matters alongside the place of human life in the cosmos. Seneca's observation that wind's unpredictable qualities make it an elusive subject has doubtless stood the test of time.[14]

Francis Bacon's *History of Winds* is widely viewed as the first comprehensive study of wind to appear in the Western world since Aristotle's *Meteorologica*. It is worth noting the historical context of Bacon's significant contribution. As in the classical world, wind proved a divisive and enigmatic subject in scholarly discourses of the Renaissance. Although prevalent in most university settings, Aristotle's ideas were not universally accepted. An important diversion was offered in the work of the Spanish naturalist and missionary José de Acosta, whose *Natural and Moral History of the Indies* of 1590 focused on personal observations and the value of experimentation, offering a detailed analysis of the trade winds, and the connection of wind to landscape and climate. Of all the early modern authors, Acosta's contributions are those most cited by Bacon in his *History of Winds*.[15]

'Francis Bacon', frontispiece to Francis Bacon and William Rawley, *Sylva Sylvarum* (1683 edn).

The title of Bacon's work *History of Winds* may suggest a backward-looking treatise on the subject, but while Bacon engaged energetically with the histories of wind that preceded him, his work was intended to provide a solid foundation for new studies of the topic in the future. Bacon gathered what information was known about winds, categorized different types of wind in terms of qualities such as seasonality ('Every wind has its weather') and projected future paths for wind's scientific study. Bacon's use of experiments and observations, his rejection of the prevailing views of Aristotle and his emphasis on greater empiricism in the study of wind had a lasting impact on the trajectory of weather science.

Early wind observations

Wind is a ubiquitous presence on the Earth's surface, but its intangibility and dynamism make it a challenging subject for scientific study. Until very recently, scientists had limited options for measuring wind. Devices used to measure wind speed are commonly called anemometers. This term comes from the Greek words *anemo*, meaning 'wind', and *métron*, which relates to measurement. As is clear from the broad meaning of the term, the word 'anemometer' can be used to describe an assortment of devices.

The first known anemometer was a mechanical instrument invented by the Renaissance artist and cryptographer Leon Battista Alberti in Italy around 1450.[16] Alberti's anemometer was crafted from a disc that hung downwards at a perpendicular angle from a stand. In still weather, the disc remained inert. When the wind picked up, the disc was deflected. The angle of the disc's deflection was decided by the speed of the wind. Alberti's device was further developed in subsequent years, with some alterations credited to well-known innovators. Leonardo da Vinci is known for his prescient inventions and fascination with developing the means for human flight. These interests overlapped in the development of the anemometer. In the 1480s, the young Leonardo made numerous alterations to Alberti's wind-measuring design during his study of flight.[17] Perhaps on account of his fame, and his association with innovation and flight, Leonardo da Vinci is widely credited with building the first anemometer in the modern day.

The confluence of wind measurement and human flight that so intrigued Leonardo has continued into modern-day meteorology. Transponders used for air traffic management, known as

Anemometer used to measure wind movement at an evaporation pan in Nevada.

Mode-s EHS, are also used to record wind data. During flight, aircraft are interrogated every four seconds by terrestrial radar systems that capture detailed information on position, air and ground speed, flight level and magnetic heading. This information is used by air traffic controllers to monitor aircraft safety, but it can also be used to track high-altitude wind speed. This is because the effects of wind resistance can be calculated through the collection of other flight data.

After Leonardo, many other inventors, including the English physicist Robert Hooke, also created new versions of the mechanical anemometer. In 1672, Hooke created what is

Leonardo da Vinci's
wind-measuring
design, fol. 675r in
Codex Atlanticus
(1486).

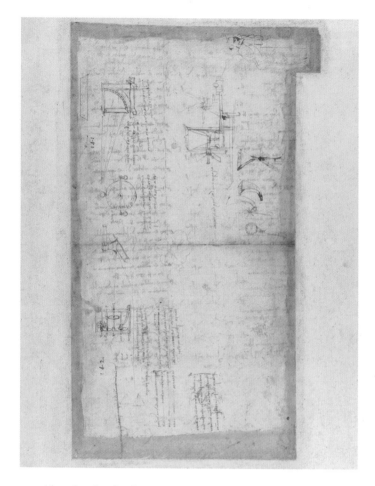

considered to be the first rotation or windmill anemometer. The most common design, with small spinning cups around a central stand, is thought to have first been developed by the Irish astronomer Thomas Romney Robinson in 1846. In an interesting turn of history, Robinson was related by his second marriage to Lucy Jane Edgeworth, whose uncle on her mother's side was none other than Francis Beaufort, the inventor of the Beaufort scale for measuring wind.

In the five hundred years that have followed the invention of Alberti's anemometer, methods of wind measurement have greatly diversified. Wind is now measured using weather balloons and weathervanes, and also satellites. In recent years, wind quantification has benefited from various technological innovations, including many in the field of space science. New radar technology, called Doppler Aerosol Wind Lidar (DAWN), offers greater detail on the movement of wind than was previously possible. To use the radar technology, researchers fly in aircraft with the radar equipment, and shoot lasers at the wind. The scattered light then bounces back at the instrument, allowing it to 'see' atmospheric aerosols, particles and molecules as the wind carries them. The aerosol particles emitted are so small that they can move at the same speed as the wind. Information gathered using this technology can inform on wind speed and wind direction (known as wind vector). This information may be used to make more accurate weather and climate predictions.

The Robinson anemometer.

Anemometer in an Austrian field.

The DAWN technology is also used to verify the accuracy of measurements provided by ADM Aeolus, a European Space Agency satellite which shoots lasers in an ultraviolet wavelength at Earth from space.

The historical process of refining and improving devices to measure the speed and power of wind has involved many creative thinkers and countless hours of scientific experimentation. From its humble genesis during the Renaissance, the anemometer has grown and diversified to provide invaluable knowledge about the natural qualities of the movement of air. While the knowledge gained from wind-measuring instruments is undoubtedly priceless, the global market for anemometers has a growing pecuniary value. By the time of writing in 2022, the market for anemometers had become worth hundreds of millions of dollars. The growth in this market has been driven by numerous forces, including the rise in consumption of wind energy, the increase in offshore wind farms and increasingly supportive regulatory policies for the growth of the wind energy market. The use of anemometers in the mining industry has also helped.

Despite the many advances in wind-measuring technology, there are many areas of wind movement that are still poorly understood. Certain types of wind and their behaviours remain difficult to predict, with the movements of low-level winds, particularly in rough terrain, proving especially resistant to efforts at neat quantification. These types of wind are frequently found around wind turbines, making them a crucial area for scientific study.

Wind records

The ability to accurately measure the power of wind in the modern day has resulted in some remarkable records. On 10 April 1996, Australia's Barrow Island recorded the fastest wind speed that has ever been measured, independent of a tornado. The wind was associated with Tropical Cyclone Olivia. The previous record was held by a wind of 371 km/h (231 mph), recorded by the Mount Washington Observatory in New Hampshire in 1934.

Sign on the Glen and Mount Washington Stage Company Office (original Mount Washington Observatory building) declaring the highest wind ever recorded, on the summit of Mount Washington in Coos County, New Hampshire.

The somewhat laconic reception for the record wind in Australia meant that it was not officially recognized as a milestone weather event until over a decade later. The director of the Mount Washington Observatory, Scot Henley, commented on the remarkable fact that it was only in 2010, after a meeting of a panel of experts from the World Meteorological Organization (WMO), that the record-breaking status of the Australian wind was noted. Mount Washington retains its record for the fastest recorded wind in the Northern and Western hemispheres. The record-breaking Australian wind was measured at 407 km/h (253 mph).

Winds in the stars

'Solar winds' are streams of charged particles moving out of the centre of the Sun (or its heliosphere). These blazing hot winds can move away into the solar system at speeds of 900 km/s (559 mi./s), at temperatures of 1 million°c (1.8 million°F). Solar wind was first discovered in 1959, shortly after the launch of the first satellites into space. The discovery is generally ascribed to Eugene Parker, an American solar astrophysicist.[18] Parker was working at the University of Chicago in 1957 when the then thirty-year-old scholar 'stumbled across' the mathematical formulas that established the presence of solar wind.[19]

Despite the significance of Parker's discovery, his ideas were not immediately accepted by the academy. A paper he submitted to the *Astrophysical Journal* was rejected by the reviewers, who suggested that he return to the library and read up on the topic before attempting another paper. The editor of the journal was the Indian-American astrophysicist Subrahmanyan Chandrasekhar, who later won the Nobel Prize in physics for his work on the evolution of stars. Chandrasekhar overruled the reviewers and published the paper. The 'radical' views of Parker posed a challenge to the traditional stance of his colleagues: that the spaces between stars were a vacuum. In 1962, however, an American robotic space probe, Mariner 2, journeyed to Venus and measured the solar wind on its way. The measurements from Mariner 2

confirmed the earlier data collected by Luna 1, which in 1959 was the first spacecraft to reach the Moon. The data collected by the probes confirmed Parker's hypothesis to be correct, ushering in a new era of understanding wind, the stars and our planet.

In recent years, scientists have observed that the temperature of solar winds reaching Earth's surface is much higher than might be expected. Indeed, data collected by satellites has found winds reaching Earth from the Sun are ten times hotter than predicted by scientific modelling. It is thought that this effect may be caused by electrons in the stream of solar wind taking longer to cool than previously predicted, due to the attractive pull of the Sun.[20]

In 2013 it was discovered that Earth produces its own version of 'space wind'.[21] The space winds (also called 'plasma-pheric winds') were shown to be travelling around the edges of Earth's atmosphere at speeds of around 5,000 km/h (3,100 mph), continuously flowing from the upper atmosphere into the

Solar winds hitting particles in Earth's atmosphere, causing the northern lights (aurora borealis), Norway.

Northern lights in Finland.

magnetosphere. The understanding of plasmapheric wind and its role in the universe is still developing: space winds are involved in numerous cosmic phenomena, including shaping entire galaxies. Plasma, gases and other matter between stars are blown through the universe by powerful winds generated by supermassive black holes. The winds that flow from black holes can change their temperature rapidly by millions of degrees, travel at a quarter of the speed of light and are sufficiently powerful to influence the formation of stars.[22]

Geomagnetic storms

Solar winds bring vital warmth to our planet, but at times the intensity of this wind can be almost too hot to handle. Usually, the solar winds of the Sun flow somewhat evenly, moving at faster and slower speeds, or in what are described as 'velocity spikes' (sharp increases in solar wind speed).[23] Occasionally, these spikes, such as those seen with a solar flare, can move extremely fast. In these events, shock waves from solar wind can upset Earth's magnetosphere (the area of space surrounding Earth, influenced by its magnetic sphere). These events are called geomagnetic storms, and they can have powerfully disruptive effects on the space surrounding Earth, as well as the planet itself.

Geomagnetic storms are best known for their deleterious effects on human technology. These powerful winds black out radars, disrupt navigation systems (such as GPS) and create harmful geomagnetic-induced currents (GICs) in power grids. As well as interrupting communications systems, geomagnetic storms affect humans and other living organisms. The changes to magnetic fields brought by the storms affect human health in numerous ways but are perhaps most clearly observed in the cardiovascular system. Increases in the intensity of solar winds have been shown to result in increased heart rates in humans, in a biological stress response.[24] Research on Russian cosmonauts has shown changes of pulse, increases in blood pressure and irregular heartbeat rhythms connected with the appearance of geomagnetic storms.[25]

Although the effects on most people are minor, there are at-risk groups of people who are more vulnerable to the storms' influence. The environmental changes may be more or less influential depending on individual sensitivity, health status and capacity to effectively regulate body systems. Those shown to be generally more vulnerable to the effects of geomagnetic storms include adults with uncontrolled hypertension and children – particularly newborns – who have yet to fully develop their body's adaptive systems to changes in the environment. The potential hazards caused by geomagnetic storms and solar winds mean that the ability to accurately forecast them is extremely valuable.

The Moon, the wind and wind beyond the solar system

It is well established that the Moon and the wind both exert an effect on the movement of water on Earth, particularly in ocean tides. Less well known is the idea that space wind may also play a role in the Moon's atmosphere. Recent studies have shown that the lunar surface is not entirely dry, even on its sunlit surfaces. Scientists are exploring the theory that hydrogen, carried to the Moon's surface by solar winds, may play a role in the creation of lunar water.

In coming years, it is likely that scholars will further expand our knowledge of extraterrestrial wind – beyond the stars, the Moon and the Sun. In 2020 an international team of scientists was able to measure winds occurring on a brown dwarf, catchily named '2MASS J1047+21'. A brown dwarf is a stellar object with a mass between the size of a large planet and a small star. As brown dwarves do not have sufficient mass to create much light, they are sometimes (rather harshly) labelled as 'failed stars' by astrophysicists. The measuring of wind speed on the brown dwarf 2MASS J1047+21 marked the first-ever wind speed measurement from beyond our solar system. Using a combination of infrared emissions and radio waves, the team of scientists detected winds travelling at incredible speeds on the brown dwarf – the recorded speeds reached 650 m/s (2,100 ft/s), with the measured winds moving in a west-to-east direction.[26] The scientists concluded that their techniques may soon allow them to measure the wind speed on other stellar objects, such as exoplanets (planets outside our solar system).

Wind and water

Moving closer to home, the influence of wind is found in the movement of currents of water on Earth. The Antarctic Circumpolar Current, or ACC, is the world's strongest ocean current, extending from the bottom of the ocean to the sea surface, and encircling Antarctica. The ACC is created by strong westerly winds blowing across the Southern Ocean and the changing surface temperatures between the equator and the poles. By keeping Antarctica cool and frozen, the wind-generated Antarctic Circumpolar Current plays a vital role in the global circulation of water and the health of the planet.[27]

Antarctica.

Sky view of Earth, with winds, tides and density differences stirring the oceans. Tropical Cyclone Joalane, which developed in April 2015, can be seen over the Indian Ocean.

In recent years, the enormous impact of wind on the ocean has begun to win greater recognition. As well as influencing the surface of the water in the movements of tides and waves, wind is also at work beneath the surface of the ocean. Wind exerts a powerful effect on the movement of currents in the water and on the composition of the global atmosphere. The study of coral fossils has recently shown that ocean circulation, even in deep water, is remarkably sensitive to wind changes. The carbon levels of the deep-water fossil skeletons from the Southern Ocean, near Antarctica, were found to appear similar to those of their counterparts in the Pacific Ocean. The discrepancy between the movements of water masses and expected circulation patterns was attributed to the influence of surface winds.[28] The dynamic effect of wind on ocean currents means that it has a flow-on effect on the release of carbon dioxide stored in deep levels of the ocean into the atmosphere. Future studies will explore how the sensitivity of water currents to wind may effect projections for climate change.

Aeolian processes

Aeolian processes describe the power of wind to shape planets. The name 'aeolian' comes from the Greek deity Aeolus, the keeper of the winds, who appears in Homer's epic the *Odyssey*.

The terraforming powers of wind are perhaps most often thought to be associated with erosion, but aeolian processes also involve the transportation and deposition of materials: as well as wearing down the earth's surface, wind builds it up. Erosion by wind occurs through the process of deblation, in which loose materials are lifted away by winds; through abrasion, in which surfaces are ground down through the action of wind; and by sand-blasting from wind-born particles. The movements of glaciers and water have traditionally been considered as the world's most powerful geo-shaping forces, but recent research has shown that wind is equally significant in influencing Earth's physical form. Scientists from the University of Arizona studied gigantic wind-formed ridges of rock called 'yardangs' to discover that the eroding power of wind is sufficient to keep mountains from growing.[29]

As well as building up and smoothing down the planet above the ground, wind erosion has been found to work on the shape of things beneath the surface. Wind erosion is involved in tectonics – the forces that cause the movement and bending of Earth's plates. Through the wind's movement of sediments, the rate at which bedrock is laid down may be accelerated or altered, ultimately effecting the underlying progression of the tectonic plates.[30] The interplay between the movement of wind and tectonic forces has also been observed by researchers working on the formation of the Andes Mountains in Argentina.[31] It is thought that similar connections between wind erosion and tectonics may be found on our windy neighbouring planet, Mars.

Back on Earth, many of the wind's eroding patterns form part of seasonal weather fluctuations, travelling from one side of the globe to the other. An example of this is the passage of dust from the Sahara, which is carried as far as the Amazon basin and the islands of the Caribbean. Scientists from NASA have shown that phosphorus in the Saharan dust provides nutrients for plants growing in the Amazon rainforest.[32] The movement of dust along wind currents has been shown to be highly variable, relating to factors such as climate and rainfall in the desert. Further from home, aeolian processes have recently been observed on Titan, the largest moon of Saturn, and Mars. The evidence for wind erosion on planets beyond Earth includes the appearance of yardangs – similar in appearance to those found in the deserts of Arizona.

As it gradually shapes new landscapes, wind erosion moulds natural wonders from rocky terrains that become part of the location's distinctive identity. A famous example of the natural designs forged by wind is the landform known as 'The Wave' in the Navajo sandstone rocks of the Vermilion Cliffs National Monument in Arizona. The swirling striped patterns and wave-like shape of the rock are the product of 190 million years of wind erosion and have made the area a popular destination for ecotourism. Although wind erosion usually takes hundreds of millions of years to craft landforms into new and, at times, wondrous shapes, sometimes the process can be remarkably rapid. Several famous rocks rising from the sea surrounding the Isle of Wight in the English Channel have been broken off by powerful winds in recent years, dramatically altering the landscape.

The power of wind to carry nutrients across vast distances fuels the ecosystems of the areas where these resources are deposited – but it can also be damaging and depleting for the areas from which the nutrients are drawn. Wind erosion is generally divided into three categories: surface creep, saltation and suspension. The categories are determined by the size of particles being transported by wind, and the type of movement involved. Human activities can exacerbate or mitigate the harmful impacts of wind erosion. Planting trees and bushes and protecting topsoil

puts obstacles in the way of wind stripping the nutrients from a region. By contrast, deforestation and overgrazing by livestock are examples of activities that can intensify the wind's effects.

The Wave, Arizona.

Sound propagation and wind gradient

Along with nutrients, seeds, spores and pollen, wind plays a crucial role in carrying sound. Through a physical phenomenon related to refraction (the bending of a sound wave owing to changes in its speed), sound waves travel more quickly in the air when moving with the wind. Sounds moving with the wind can be heard over longer distances. The amount of influence that the wind will hold over the distribution and quality of a sound is determined by the 'wind gradient'. The 'wind gradient' is the vertical gradient of the mean horizontal wind speed in the lower atmosphere.

The concept of wind gradients has wider applications than the movement of sound and communication; it is also important in human activities such as gliding and sailing. As well as carrying sound, the wind gradient is utilized by large birds to enable flight. Using high aspect ratio wings, birds can soar high above the ocean. In a manoeuvre known as 'dynamic soaring', large birds such as albatrosses can gain energy from the wind to power sustained, non-flapping flight.

It would appear that the wind gradient, along with thermal air, has been harnessed for flight for a very long time – palaeontologists working on the giant ancient pterosaur Quetzalcoatlus have theorized that the dinosaur used the wind gradient and its 10-metre (34 ft) wingspan to lift itself high above the terrain.[33] The dinosaur is thought to be the largest flying animal ever discovered, and its large size meant that using the wind to soar was particularly critical for successful flight. As an animal's body size increases, the dynamic power of its movements decreases, meaning that flapping movements become less useful for remaining

The cliffs of the Isle of Wight.

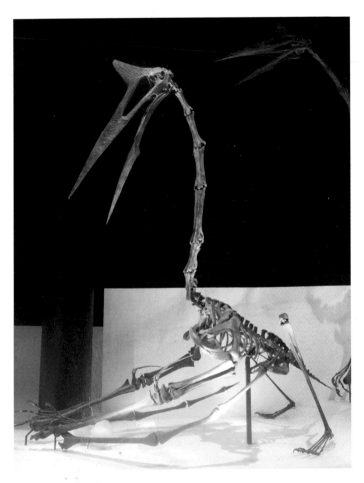

Quetzalcoatlus fossil in the Houston Museum of Natural Science.

airborne. For larger animals, such as flying dinosaurs, soaring on the wind makes flight possible.

Talking trees and whispering winds

It is not only human and animal communications that are carried by the wind – the movement of air forms a kind of natural instant messaging system for the world's plants. Research has shown that trees and other plants are able to communicate with one another using airborne gases that are carried from one organism to the other by wind. The gases used by the plants to communicate are called 'volatile organic compounds', which are perhaps more commonly known as perfumes emitted by flowers, or the sweet smell of freshly mown grass. In the case of cut grass, the damaged plant emits a type of 'distress signal' compound, which alerts other plants nearby of the threat posed by a lurking lawnmower. The deck may be heavily stacked against the grass in its battle with the lawnmower, but other plants have used their windborne messaging systems to protect themselves from predators.

An example of this 'wood-wide web' has been found on the savannas of sub-Saharan Africa. The umbrella thorn acacia tree is an emblematic species in the region. Along with its character-istic thorns, the tree uses subtler (yet no less dangerous) tools to ensure its survival. When an animal, such as an antelope, begins to chew on the leaves of the acacia tree, it emits a quantity of ethylene gas into the air.[34] The gas is carried on the wind to other acacia trees in the area, who respond to the message by increasing the tannins in their leaves. These tannins are toxic and may even prove fatal to the large mammals that graze on the acacias' leaves.

Interestingly, giraffes have shown an impressive ability to use wind to counter the defensive messaging of the African acacias. Giraffes tend to graze downwind, preventing the trees from using their messenger system to alert their tree neighbours. In this way, the giraffes mirror the strategies of carnivorous animals who use the wind's direction to mask their presence when stalking prey. If there is no wind, the giraffes spread their grazing out further than the ethylene gas can be carried, again circumnavigating the trees' defences.

Wind fertilization

This creative role of wind is frequently reflected in myths and legends of many cultures around the world, such as in the Greek myth of Chloris, a flower nymph abducted by Zephyrus, the west wind deity. As well as carrying seeds and distributing flowers and other plants, wind is also involved in the movements of airborne pollinating creatures.

Pollination is the process of pollen grains moving from a male part of a plant to a female part of a plant, allowing for fertilization to occur and seeds of new life to be generated. Wind is one of the world's most important pollinators, filling the role of transporting pollens alongside birds, butterflies, bees, water and a host of other creatures – even turtles. Interestingly, the different types of natural pollinators tend to work on different types of plants. For example, water-pollinated plants tend to be aquatic, and insect-pollinated plants tend to be large and brightly coloured, so as to attract the pollinator to the plant. Wind-pollinated plants are said to be 'anemophilous' (wind-loving) and tend to be smaller and rather drab in appearance when compared to their flashier insect-pollinated neighbours. It is thought that around 12 per cent of flowering plants are anemophilous, with well-known examples being oaks, alders, pecans, pistachios and many species of fir, pine and spruce trees. Despite their less colourful appearance, anemophilous plants do not go entirely unnoticed by humans – pollens from wind-fertilized plants are most commonly to blame for causing seasonal allergies such as hay fever.

Some plants have adaptations to assist the process of wind dispersal, such as 'winged' maple tree leaves and the light, feathery seeds of cottonwood trees and dandelions which make them easier to carry. Other plants' seeds, such as those of milkweed, resemble tiny parachutes. Once a plant has taken root, wind

Milkweed seeds ready to fly on the wind at Minnesota Valley National Wildlife Refuge.

influences every part of its development, including the organism's chemical composition, its physical structure and morphology – from the cellular level to the plant as a whole.[35] Beneath the soil, the root systems of plants are influenced by the types of wind blowing above the surface, as powerful winds necessitate deeper, sturdier root systems to prevent uprooting. Above the ground, plants must adapt not only to the force of the wind, but to wind's constant fluctuations – the response of plants to wind is dynamic and constantly adapting to the environmental changes that wind creates.

Tree with extensive root system.

As well as assisting plants to grow through the process of wind-assisted fertilization, or 'anemochory', wind can also limit the heights to which the new plant growths will climb. Strong winds can stunt the growth of trees, resulting in smaller trees in wind-exposed regions such as coastlines. The natural movement of wind plays a vital role in the development and maintenance of healthy forests through 'wind pruning'. This process sees the removal by wind of dead, damaged or structurally unsound branches from trees in order to increase the plant's resilience.

The shaping relationship between the wind and plants is not one-sided. Recent studies have shown the capacity for plants to change the velocity of wind speed and reduce aeolian (wind-caused) erosion.[36] Plants provide wind with a voice, through leaf flutter and other aeolian sounds (sounds created by wind passing an obstacle). The movement of plants gives a 'visual manifestation of the reality of wind'.[37] Plants further provide humans with a means of quantifying the unseen force of wind, through the empirical measure of the Beaufort scale. At level 10 on the Beaufort scale, winds are sufficiently powerful to uproot trees – an occurrence known as 'windthrow'.

Wind may, at times, assist other pollinators – monarch butterflies are known to make use of gusty tail winds in their 2,500-kilometre (1,500 mi.) flight from Canada to Mexico. At other times, however, wind may prove an obstacle to fertilizing creatures. Recent scientific modelling has shown that the United Kingdom is likely to see a large increase in frequent and intense wind storms over the next few decades, due to the effects of global warming.[38] Experiments with the flight patterns of bees in low or high winds has shown that high winds greatly reduce the effectiveness of bees in their pollinating missions.[39] Even tiny increases in wind speed reduce the body temperature of the bees, meaning that it takes longer for them to warm up their flight muscles. The researchers also speculated that the bees may wait for a break in the wind to take off, a behaviour that has been observed in albatrosses.[40] With blustery conditions predicted to increase with climate change, insect pollinators may be faced with a significant impediment to their fertilizing efforts.

Forest at Lake
Superior.

While wind plays a creative role in spreading plants and assisting in promoting their growth and energy production, wind damage to plants has significant ecological and economic costs. Wind accounts for over 50 per cent of all damage by volume to forests in Europe.[41] In 1999 two cyclones in western and central Europe blew down 176 million cubic metres (more than 6 billion cubic feet) of wood valued at €6–7 billion.[42] Globally, wind

Uprooted redwood
tree, Redwood
National and State
Parks, California.

damage results in the widespread spoilage of agricultural crops
such as cereals, fruits and vegetables, with heavy societal and
economic costs. Damage may include the uprooting of plants,
tearing of leaves or the abrasive damage of soil and sand blowing across crops. Wind also causes indirect damage to plants by
transmitting diseases, such as wheat and coffee rusts, or pests,
such as locusts. Wind damage to agriculture is not limited to
plants – fisheries, for example, may experience decreased oxygen-
ation of water pools in high winds, leading to fish death events.
While wind damage can be reduced through actions such as
creating shelters and windbreaks, and through strategic planting, the unpredictability of wind can make challenging work of
preventative measures.

From the formation and distributions of stars, to the travel
patterns of monarch butterflies, wind plays an unseen yet vital
role in shaping the cosmos. In the exhalations of supermassive
black holes millions of kilometres away, or in the shifting of tectonic plates beneath our feet, life on Earth is shaped by wind at
every level. Understanding and defining this force and its effects
has puzzled great minds throughout the course of civilization.
Indeed, efforts to comprehend wind have played an important

Migrating monarch butterflies (*Danaus plexippus plexippus*) on a cedar elm tree in central Texas.

role in fuelling the growth of numerous scientific disciplines, discoveries and inventions.

Wind by its very nature is always shifting, and in recent years the changes to global wind patterns through climate change have signalled even greater dynamism ahead. What remains certain in this fluctuating landscape is that the invisible touch of wind will continue to be felt – carrying life and its materials to the furthest pockets of the galaxy.

2 Wind in the World's Oldest Literature

In some of the world's most ancient literature, wind is created by deities, weaponized by heroes and invoked in magic. Wind is presented as a creative and destructive power, but it is not considered invulnerable – at times, wind can itself be injured through human error. Distinctive winds feature in Egyptian myths, magic and literature, including the Book of the Dead, and the four winds are connected to resurrection. Winds also appear prominently in many Mesopotamian legends, including the *Epic of Gilgamesh*, *Adapa and the South Wind* and the Babylonian creation story, *Enuma Elish*. Wind is tied to creation and wisdom in the Hebrew Bible, from the Book of Genesis to the Song of Songs.

Wind is a truly universal phenomenon, yet it is conceptualized distinctively across the early written cultures of the ancient Near East. At the same time, the connection of wind to religion in some of the world's oldest written traditions shows some intriguing areas of commonality.

Wind in ancient Near Eastern literature

The world's first known writing is found in Sumer (located in modern-day southern Iraq) around 3300 BCE. This was the time of the appearance of the cuneiform script – an intricate system of lines and wedges dug into clay tablets representing words, letters and symbols, and resembling chicken scratches in mud. For over 3,000 years, cuneiform was the primary medium of written communication in the ancient Near East, providing

Statue of the Sumerian wind god Enlil, from the Scribal Quarter at Nippur, Iraq, 1800–1600 BCE, baked clay.

an often-overlooked influence on later literature and cultures. Cuneiform literature includes texts written in Sumerian, Babylonian and Assyrian, as well as texts from the Hittites, Hurrians and others from the Levant.

The Epic of Gilgamesh

In literature from ancient Mesopotamia (an area roughly contiguous with modern-day Iraq), a trend appears that continues throughout much later literature – the fusion of wind and religion in human thought. Several Mesopotamian deities have connections to wind. Enlil is a primary deity in the Mesopotamian pantheon, situated at the top of the divine hierarchy from the earliest times. Enlil appears in the most famous Mesopotamian literary masterpiece, *The Epic of Gilgamesh*. In the epic, he attempts to destroy humankind with a giant storm, in a story often paralleled with the narrative of Noah's Ark in the biblical Book of Genesis.

Enlil is described as 'king' or 'supreme lord', but his name literally means 'Lord Wind'. Enlil's wife, Ninlil, has the complementary name meaning 'Lady Wind'. Legendary historian Thorkild Jacobsen described Enlil as the 'personification of the wind and the numinous power in it'.[1] The nature of Enlil's association with wind changed over time, from being a manifestation of moving air, through being the breath issuing from his mouth – notably his 'word' or 'command'. As the embodiment of Enlil's word, wind carried forth divine judgements to the rest of the cosmos. The connection of wind with powerful words and breath is also found in Egyptian myths. Enlil's connection to wind in Mesopotamian myth reflects both the creative and destructive aspects of the natural phenomenon. Enlil is connected to powerful and destructive storms, but also to reviving winds bringing rain and greenery to deserts and fields. In the *Epic of Gilgamesh*, Enlil's use of wind reveals his capacity for destruction.

The *Epic of Gilgamesh* tells the story of the young hero Gilgamesh and his journey towards accepting his mortality and his role as king of Uruk. It is the world's oldest known epic, with

Cuneiform tablet
with Sumerian
inscription from Ekur,
the temple of the god
Enlil, *c.* 16th–15th
century BCE.

versions first written down around 2100 BCE. Towards the end
of the *Epic of Gilgamesh,* legendary flood survivor Utanapishtim
tells the hero about when the gods (led by Enlil) created a
plan to destroy humanity. This section of *Gilgamesh* has been
the focus of a great deal of scholarship, due to its numerous
close parallels with the biblical Flood narrative of the Book of
Genesis. Along with the many similarities between the flood
narratives of *Gilgamesh* and Genesis, several differences have
been noted. One primary difference relates to the description
of the destructive weather event itself. The biblical account is
sparser than that in the Mesopotamian myth, with a more singu-
lar focus on water compared with other weather features. Water
is significant in *Gilgamesh* also, with a narrative focus on the
building of a boat. There are, however, references to other types
of weather accompanying the rising waters, including winds,
gales and 'torches' in the sky held by deities – possibly a reference

to lightning. In *Gilgamesh*, the destructive force of the storm is said to frighten even the gods, who retreat to the highest part of heaven – again showing a contrast with the biblical narrative. Whereas wind plays a more visible role in the destructive flood from Mesopotamia, in the biblical Flood, wind is used to save all living things from the rising tides.

Enuma Elish

The Babylonian creation myth, *Enuma Elish*, depicts a battle involving an array of savage winds. Records of the myth have been discovered dating to 1200 BCE, and these are likely later editions of even earlier versions of the story. The myth gives an account of the rise of Marduk, the chief Babylonian deity, to the top of the divine pantheon. The personified primordial ocean, Tiamat, engages in a stormy battle with Marduk that reshapes the cosmos. Marduk's victory over Tiamat sees him rise to power among the gods, and ensures the continuation of universal order. Fundamental to Marduk's epic win is his use of the four winds: North, South, East and West. The cardinal winds are newly created in the story and given to Marduk by his sky deity father, Anu. Marduk looses the four winds upon Tiamat, and they drive her into a state of agitated confusion and distress. As Marduk bears down on his enemy, he again sets the winds to surround the primordial creature, encircling her and cutting off her escape.

Along with the cardinal winds, Marduk creates seven powerful new winds to assist in his battle (creating winds seems to be a divine family trait). Among these are the hurricane, the 'evil wind'; the dust storm, the chaos-spreading wind; the 'four-fold wind'; the 'seven-fold wind'; and the cyclone. After some preliminaries, Marduk sets the 'evil wind' on Tiamat, who attempts to swallow it. The young god then shoots Tiamat with an arrow, and she dies. Marduk sends the North Wind as a messenger to the other deities to share news of his victory.

Top fragment of a *kudurru* with a *mushhushshu* dragon and divine symbols. The dragon and the spade are symbols associated with the deity Marduk. The *kudurru* dates to the second dynasty of Isin, *c.* 1156–1025 BCE.

Ishtar

Enlil and Marduk are not the only Mesopotamian deities to have harnessed the destructive power of wind. The phenomenon was used to carry out the will of Ishtar (Inanna in Sumerian), a complex goddess associated with love, war and social connections.[2] Ishtar was, at times, said to live in 'the house of the winds', and she uses wind in myths as a weapon of war and mass destruction, with a common element of these myths being the goddess's pursuit of vengeance.

Compositions by the ancient princess and poet Enheduanna describe Ishtar's ability to use winds in her favour. Enheduanna was a historical figure, known as the world's first individually

identified author. In one myth written by Enheduanna, *Inanna and Ebih*, Inanna (or Ishtar) goes to war against a mountain (the eponymous Ebih). The cause of friction between goddess and landmark is a perceived lack of reverence from the mountain, who refuses to bow low to the goddess. Inanna stirs up a raging wind to assist in her preparations for battle. The tempestuous love goddess again sends fierce winds against her adversary (a type of noxious weed) in the Sumerian myth of *The Shumunda Grass*, a composition of unknown authorship. Once more showing her preference for settling scores through atmospheric means, in the myth of *Inanna and Shukaletuda*, the love goddess sends winds and a dust storm across the land. The bad weather features as one of a series of plagues following her rape by the gardener, Shukaletuda, with the other plagues involving a river of blood and a traffic jam. A hymn written by Enheduanna describes the goddess as 'clothed in a furious storm, a whirlwind', and in other texts she is portrayed as riding on top of the South Wind.

Panel with striding lion, *c.* 604–562 BCE, Babylonian. The lion is an emblematic animal of the primary Mesopotamian goddess, Ishtar.

Other mythical references to wind

Winds in Mesopotamia were conceptually related to supernatural beings and activities, as well as to magic. Among a variety
of supernatural beings related to wind in Mesopotamia are three
wind demons, *lillu*, *lilitu* and *ardat lili*. The three often occur
together, and they are related to stormy winds. It is thought that
'Lilith', the dangerous demon from Jewish magical folklore, takes
her name from Sumerian wind demons, although the connection is disputed in scholarship. Like Lilith, the Sumerian wind
demons have a sexuality linked to risky carnal activity rather than
to fertility or motherhood.[3] The *Sumerian King List* (an ancient
document listing historical and legendary kings of Sumer and
neighbouring areas) states that Gilgamesh's father was a *lillu*, or
a phantom – suggesting that not all sexual activity relating to
Sumerian wind-spirits resulted in barrenness. The connection
between wind and fertility also features in classical myth, considered in the next chapter. Wind was used in Mesopotamian
magic, notably in rituals to contact the dead. The first funeral
rite among those performed over the dead during rituals of the
afterlife was the ceremony to 'blow away the wind'. This ceremony was crucial for the well-being of the person's afterlife, and
it functioned to separate, or 'loosen', the spirit from the body.[4]

In the modern day, wind is widely recognized for its
capacity to take a variety of forms. There are breezes and north-
easters, hurricanes and gales, twisters and crosswinds. Ancient
Mesopotamians also recognized a number of types of wind –
indeed, this is a common feature of human responses to wind
across most ancient cultures. The four winds, named after the
four main cardinal directions (north, south, east and west), feature in the Babylonian story *Enuma Elish* but also in myths,
proverbs, royal hymns and legendary stories of ancient heroes
found in Sumerian literature. As with many cultures, winds from
different directions were viewed as holding distinctive attrib-
utes. A Sumerian proverb, dating to the third millennium BCE,
distinguishes the innate qualities of the four winds:

The north wind is a satisfying wind; the south wind is harm-ful (?) to man. The east wind is a rain-bearing wind; the west wind is greater than those who live there. The east wind is a wind of prosperity . . .[5]

While the winds from each direction were powerful, the South Wind was particularly fearsome. The destructive and often malign power of the South Wind is shown in its frequent description as an 'evil wind'.

Adapa and the South Wind

Despite its ferocious reputation, the South Wind manages to run into trouble in the Mesopotamian myth of *Adapa and the South Wind*. In this Akkadian legend, Adapa is a mortal who is given perfect wisdom by Ea, the Mesopotamian god of wisdom. Although usually zealously pious, Adapa becomes angry when the wind capsizes his boat while he is fishing. He reacts by

Helmet with divine figures beneath a bird with outstretched wings, dating to the Middle Elamite period, *c.* 1500–1100 BCE. The central figure is an Elamite water deity, considered to bear similarities to Ea, the Mesopotamian water and wisdom deity.

cursing the wind and fracturing its 'wing'.[6] Anu, the sky deity, summons Adapa to heaven to account for his rash actions. Ea instructs Adapa on how to conduct himself: he must display mourning behaviour to earn the good graces of the gods at the door, Tammuz and Gizzida, and refuse the food and water of death (which would kill him). Ea's advice helps Adapa to impress Anu, who then offers him the divine food and water of life. Taking the offered refreshments would make Adapa a deity, releasing him from Ea's service. Adapa, however, is guided by Ea's instructions not to eat or drink in heaven, and so he rejects Anu's offer and returns to earth a mortal.

There is no scholarly consensus on the purpose and meanings of this narrative, although the myth is certainly concerned with exploring the distinction between humans and deities, especially in terms of immortality and wisdom. Scholars have argued that Ea may have known in advance that Adapa would be offered the secret to eternal life in heaven and may have deliberately misled his servant as punishment for fracturing the 'wing' of the wind.

Broken wings

Interestingly, a hero who breaks the wind's 'wing' is also found in the legends of the Wabanaki community, a confederacy of First Nations people from North America. The Wabanaki hero Glooskap captures the Great Wind-Bird (Wuchowsen), in response to its disruptive behaviour. Glooskap in other narratives warns people of a great flood. The hero is also famous for distributing knowledge among his community. The legend of Glooskap and the Wind-Bird contains remarkable parallels with the Adapa story:

> . . . Wuchowsen, meaning Wind-Blow . . . lives far to the North, and sits upon a great rock at the end of the sky. And it is because whenever he moves his wings the wind blows, they call him that.

When Glooskap was among men he often went out in his
canoe with bow and arrows to kill sea fowl. At one time it
was every day very windy; it grew worse; at last it blew a
tempest, and he could not go out at all. Then he said,
'Wuchowsen, the Great Bird, has done this!'[7]

Glooskap warns the Great Wind-Bird not to be careless
with the community, and instead to show compassion, saying,
'You have caused this wind and storm; it is too much. Be easier
with your wings!' The Great Bird does not heed the hero's warning, replying, 'I have been here since ancient times; in the earliest
days . . . I first moved my wings; mine was the first voice – and
I will ever move my wings as I will.'[8]

Glooskap captures the Great Bird and binds its wings, before
throwing it into a deep chasm. For several months, the people of
the area can go out in their canoes, with the waters now com-
pletely calm. In time, however, the still waters grow stagnant,
becoming so thick that Glooskap can no longer paddle his canoe.
He remembers the Great Bird and its bound wings. Glooskap
returns to the chasm, frees the bound bird and releases one of
its giant wings. From that time, it is said that the winds were
never as wild again.[9]

There are numerous differences between the Glooskap and
Adapa narratives – while both feature an over-zealous wind
being punished, the results are very different. Yet the common
elements of the stories are remarkable given the vast historical and geographical distance between the origins of the two
myths. Both stories reflect a similar conception of a divine wind
whose wing is hobbled by a sailing hero in response to interference with fishing and fishing-boats. In both stories, the initial
relief of the stilled wind's calming is followed by stagnancy that
threatens the community and environmental balance. It is worth
considering possible causes for the similarities between the two
stories. It seems unlikely that these are a result of cultural contact
in modern times – the earliest written version of the Glooskap
myth appears in a book of Passamaquoddy myths collected by
Charles Leland and published in 1884. The first translation of

Wakan-chan-cha-gha (frame drum), created by an artist from the Sioux or Dakota tribe, late 19th century. The bird holding lightning arrow symbols suggests the thunderbird.

Adapa and the South Wind appeared in 1889, in a volume on the Amarna Letters by Winckler.[10] This means that either there was some ancient story-telling connection that is likely untraceable, or the similarities grew from a common conception about the power of wind – and its remarkable vulnerability to human harm – in both cultures.

Wind in the Bible

The ubiquity of wind in the natural environment is reflected in the literature of the Bible. In the Hebrew Bible (commonly known as the Old Testament), natural forces feature prominently in well-known narratives, poetry and even legal texts. Wind plays a powerful role, with biblical wind linked to creation and fate, as well as to the power of God.

Angels restrain the four winds, leaf from an illustrated Beatus manuscript, Spanish, c. 1180.

The biblical Flood of Genesis is of course overwhelmingly focused on the element of water and its destructive capacity. Yet wind also plays a critical role in this event – it is wind that God sends to bring the Flood to an end, saving terrestrial life from extinction. In Genesis 8:1, God is described as causing a wind to blow across the face of the earth to bring the Flood to an end: 'But God remembered Noah and all the wild animals and all the domestic animals that were with him in the ark. And God made

56

a wind blow over the earth, and the waters subsided' (Genesis 8:1). After the wind is sent to bring the Flood to a halt, the earth becomes habitable once again.

Throughout the Bible, the depiction of wind reflects the ancient trend of identifying winds from different directions distinctively. The idea of personifying winds based on the points of the compass is seen in the Book of Job: 'Out of the south comes the storm, And out of the north the cold' (Job 37:9).

When described together, the cardinal winds are frequently connected to apocalyptic settings in the Bible. Depicted as under the power of God or angels (or both), the four winds are involved

Angels restrain the four winds, leaf from the Wellcome Apocalypse, *c.* 1420.

in mediating the power of life and death (Job 1:18–19, Mark 13:27, Matthew 24:31, Ezekiel 37:9). This connection is frequently represented in works of art: in the work of the German Renaissance artist Albrecht Dürer, for example, angels holding the four winds appear in the graphic project *The Apocalypse with Pictures* (1498). The image represents the biblical scene in Revelations 7:1, in which angels are described 'holding back' the four winds during apocalyptic visions. The 'holding back' of the winds symbolically reflects the staying of divine judgement prior to further cataclysmic events.

When listed individually, the cardinal winds fill less eschatological roles. While the north, south and even west wind often prove useful in the biblical texts, the east wind is frequently connected to destruction and warfare. It is described as a 'scorching' wind, used by God to send a plague of locusts in the Book of Exodus. Another wind connected to divine military power is the whirlwind. This wind is used to describe the chariot of God in the Book of Isaiah. As well as scattering the wicked, the whirlwind is used by God as a means of communication in the Book of Job.

Wind and wisdom literature

The creative capacity of wind is observed in several biblical passages, such as in the poetry of the Song of Songs: 'Awake, O north wind, And come, wind of the south; Make my garden breathe out fragrance, Let its spices be wafted abroad' (Song of Songs 4:16). The presence of wind in the Song of Songs hints at a broader connection between wind and wisdom in biblical writings and wisdom literature. In biblical scholarship, the Song of Songs is often listed among the biblical 'wisdom literature' works with books such as Job and Proverbs – although its unique content and style defy neat attempts at classification.

Like the Song of Songs, the Book of Ecclesiastes (or Qohelet) is considered to fit into the genre of 'wisdom literature', and it is also closely associated with Solomon (to whom the authorship of Song of Songs is ascribed). The main figure in the Book of Ecclesiastes, Qohelet ('one who calls or assembles'), is commonly connected with Solomon, but this 'does not imply Solomonic authorship of the material attributed to Qohelet'.[11] Through allusions to the famously wise figure of Solomon, the authors of Ecclesiastes frame the book as meditation on wisdom, in the form of a father–son lesson.[12] This structural device also features in the wisdom literature of Proverbs and is commonly found in many Near Eastern works of wisdom literature – including the *Epic of Gilgamesh*. It is generally agreed that the *Epic of Gilgamesh* has experienced some form of cultural contact with the biblical Book of Ecclesiastes, with both texts containing passages extolling the wisdom of enjoying life's simple pleasures. In Ecclesiastes, a contrast is made between the wisdom that comes from God and all other things in life – or everything 'under the sun'. Wind in Ecclesiastes is presented as one of the main contrasts to divine wisdom.

In Ecclesiastes 1, the narrator describes being sent on a path to discover all knowledge, only to conclude that the search for wisdom by mortals is 'vanity and a chasing after wind'. Both 'vanity' and 'wind' are ephemeral qualities, and indeed, the ancient Hebrew word for 'vanity', *hevel*, can also be used for something transitory, such as an exhalation. The *Epic of Gilgamesh* contains a similar observation, when the hero himself notes, 'As for man, his days are numbered, all that he ever did is but wind.'[13] In both texts, the quest for earthly gains comparable to those of the divine is described as being like wind – always just beyond human reach.

Wisdom is explored in the Book of Proverbs, in which the ephemerality of the element is referenced: 'Like clouds and wind without rain, is one who boasts of a gift never given' (Proverbs 25:19). As in Ecclesiastes, wind is used in Proverbs to contrast the difference between human and divine abilities: 'I have not learned wisdom, nor have I knowledge of the holy ones. Who has ascended to heaven and come down? Who has gathered the wind in the hollow of the hand?' (Proverbs 30:4). The intangible quality of wind is further noted in the Book of Job, where the suffering protagonist and his friends accuse one another of 'windy words'. In Job, as in other books of wisdom literature, wind gives shape to the invisible boundary between the human and divine worlds.

Wind in ancient Egypt

The prominent role of the Sun in Egyptian religion is well known in the modern day. Less well known, but equally dynamic, is the connection of wind to ancient Egyptian beliefs. Several deities in the Egyptian pantheon are related to wind, with gods connected to different types of winds in different settings.

The deity Shu, brother of Tefnut, is connected to wind and daylight. Shu plays an important role in regulating the activities

Shu kneels and raises his hands to separate the heavens and the earth, 332–30 BCE (Ptolemaic period), faience.

of the four winds. In the Coffin Texts, it is stated that the wind is the *ba* of Shu, as light is the *ba* of Ra. *Ba* in the Egyptian cosmic worldview is a complex topic: it can represent a kind of soul in humans, one that leaves the body to inhabit the celestial sphere. In the divine world, *ba* conceptually linked a deity to its manifestation. The creative potency of wind is seen in Shu's role in birth and pregnancy – she is said to bring the life-force of wind to the unborn in the womb.[14] The role of the wind as a divine breath that acts as a life-giving force is found in magic spells, including in the Book of the Dead. The solar deity Amun was also linked to wind, as he was unseen, yet ever present. The lion-headed goddess, Sekhmet, has a connection to wind that reflects the dualism seen in her role with magic and medicine. Sekhmet is associated with disease and pestilence, but also with healing and curing. Similarly, while the goddess was related to scorching desert winds, she was also thought to bring forth cooling breezes.

Winds from the north, south, east and west were each part of Egyptian religion. From the time of the Pyramid texts, the four winds were viewed as servants of the Egyptian god of creation and the Sun, Ra. The four winds were thought to stand behind him, and the winds' power of 'looking with two faces' meant that their gaze could have a positive or negative effect. In the Coffin Texts, the universal quality of wind is emphasized in the story of the four winds' creation. Ra is said to have created wind (along with water and earth) so that 'the humble might benefit from (them) like the great' – an action that established the principle of equality among humans in accordance with the rule of Maat.[15] The deity Maat in Egyptian religion was the embodiment of truth, justice and moral integrity, and she represented universal order in Egypt. The four winds were personified as winged creatures who embodied the four cardinal directions. Their winged form meant they were also connected to birds and flight. Qebui was the god of the north wind, depicted as a ram with four heads and wings. Henkhisesui was the east wind, who could be shown as a winged scarab, and Hutchai was the west wind, shown as a man with a serpent's head. The last of the four was the lion-headed south wind, Shehbui. In the Book of the

Dead, the north wind is singled out as the 'refreshing' wind and the one that accompanies the *ba* of the deceased person back from the 'horizon' of the tomb.[16]

The four winds played a complicated role in ancient Egyptian magic, relating to life and resurrection, and bringing life-giving breath. The north wind is often used synonymously with breath, and the terms 'north wind' and 'sweet breath' were designations for the royal word across a range of periods and dynasties.[17] As much as they were strongly connected with life, the four winds were also important in the Egyptian afterlife. The Coffin Texts reflect the role of the four winds under the authority of the deity Shu, 'representing the totality of the divine forces that enlivens one in the afterlife'.[18] Even in death, the creative potential of wind was recognized in Egyptian religion.

Scene from the Book of the Dead for the Chantress of Amun, Nany, showing Nany before Osiris, Isis and Nephthys, *c.* 1050 BCE, papyrus.

Mysterious and magical, wind connected the divine and human spheres in Mesopotamian, Egyptian and biblical literature. The ancient recognition of wind's creative and destructive potential is seen in its presence in Flood narratives and Creation stories, spanning wide cultural and geographic boundaries.

In some of the world's most ancient literary traditions, the invisible force of wind is brought to life in diverse written texts. In the ancient Near East, the complexity of wind's role in the natural world is mirrored in the recognition of the different roles and characteristics of numerous types of wind – from whirlwinds to westerlies. At the same time, areas of cultural crossover in ancient stories of wind from different regions and historical contexts hint that some aspects of human responses to the natural world may be as universal as wind itself.

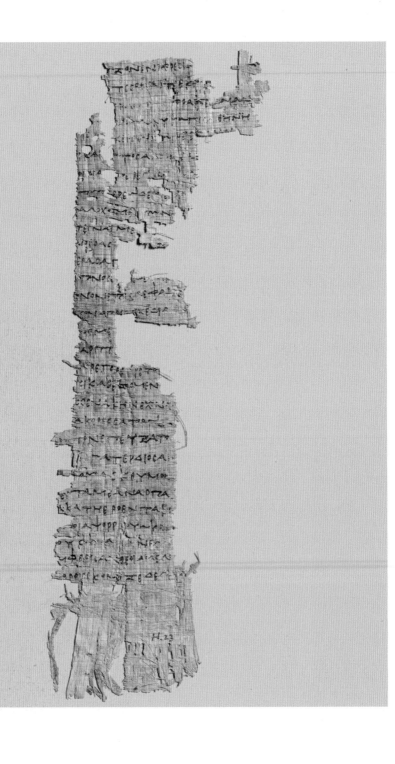

3 Myth, Folklore and Religion

In the myths, legends and folklore of global cultures, wind formed an invisible nexus between the natural and supernatural worlds. This connection is found in musical instruments, such as the aeolian harp and the famous reed pipe of the Greek deity Pan. As well as carrying music and voices, wind has long been recognized for its subtle but vital connection to breath. In the classical world, the Anemoi were four wind deities who feature in numerous myths, with the Greek word *ane* relating to breath. The connection between wind, breath and life (or 'spirit') is found in diverse cultures throughout history. In Inuit religion, for example, the word *sila* can mean 'winds', 'consciousness' or 'life-giving element'.[1] From Homer's *Odyssey* to the modern day, wind has also played a prominent role in myths involving seafaring. An example of the bond between wind and sailing from Polynesian culture is found in the story of the wind deity Paka'a, who appears in Hawaiian myth as the inventor of the sail.

Wind is a constant and dynamic presence in world myth, legend and folklore. The shape taken by wind in stories from different cultures shows remarkable diversity, yet frequently reflects its natural physical qualities.

Wind in classical mythology

Wind's role in powering the sails of seafaring vessels has shaped the distribution of trade, culture and human settlement. Its connection with seafaring has seen wind play a central role in many myths related to sailing. One of the most famed stories of a (very) long sea voyage from the classical world is Homer's *Odyssey*. Written around the eighth century BCE, the story describes the journey of the Greek hero Odysseus, as he tries to make his way home across the seas following the Trojan War (an event described in Homer's *Iliad*). Odysseus and his doomed sailing companions show impiety to the Greek god of the sea, Poseidon, whose wrath ensures their journey home to Ithaca is long and hazardous.

The *Odyssey* is noted for its complex use of wind imagery.[2] Winds in the *Odyssey* relate to time and fate: as noted by classical scholar Alex Purves, the paths of the Greek heroes homewards are shaped by the gods' control of the winds.[3] Purves's observation of the narrative tension surrounding winds, gods and control is illustrated at numerous points in the story. Poseidon seizes control of Odysseus' fate by manipulating wind, and Odysseus equally tries to regain power over his journey through avoidance of wind, and through his attempts to harness its power.

The use of wind to enact divine punishment in the *Odyssey* is echoed in a much later work, the *Rime of the Ancient Mariner* by Samuel Coleridge (1834). Similarities between the two works have often been noted in scholarship, emphasizing the capacity of wind in narrative to blend the natural with the numinous. In the poem, an albatross flies in on the south wind each day to visit the Mariner's vessel; the Mariner commits the shocking crime of killing this bird who has befriended him. As a result, the Mariner and his crew are cursed until they can atone for the crime, and justice has been served. Part of the curse is an absence of wind. Through divine intervention, the element that had initially carried the albatross to the Mariner deserts him, leaving the crew stuck in the doldrums. In both ancient and more modern narrative poems, wind is an instrument for enacting divine justice, but also an unpredictable element of the marine environment, representing an easy fusion of natural and supernatural qualities.

Returning to Book Ten of the *Odyssey*, Odysseus and his fellow Greeks sail from the land of the giant Cyclopes to the home of Aeolus, the legendary keeper of the winds. Moved by

Isaac Moillon, *Aeolus Giving the Winds to Odysseus*, 17th century, oil on canvas.

Odysseus' plight, Aeolus gives him a bag containing the many winds. Aeolus provides the weary sailor with a westerly wind to steer the lost sailors safely home. Aeolus' kind act is undone by Odysseus' crew, who open the bag of winds while the hero is sleeping, releasing a catastrophic storm that drives them straight back to Aeolus' island. The wind-keeper refuses to help the sailors a second time, as he sees their misfortune as a sign of divine displeasure.

The keeping of winds to assist in sea voyages, seen in the *Odyssey*, is a theme found in several later traditions. Practitioners of magic and sorcery in the ancient world were thought to be able to harness winds and keep them at hand by tying special

François Boucher, *Juno Asking Aeolus to Release the Winds*, 1769, oil on canvas.

J.M.W. Turner,
*Thomson's Aeolian
Harp*, 1809, oil on
canvas.

knots. The knots could be sequentially released to provide the required wind. The knot-tying tradition has also been observed among wind-harnessing fishermen in the Shetland Islands.

In contrast to his helpful role in the *Odyssey*, the wind-keeping Aeolus is the means through which the hero is blown off course in Virgil's *Aeneid*. Written around 19–29 BCE, the *Aeneid* describes the journeys of Aeneas, the Trojan ancestor of the Roman people. In Book One of the epic, Aeolus sends a violent storm that destroys Aeneas' fleet at the request of the goddess Juno. In an intriguing reversal of roles, the Roman sea deity Neptune (Poseidon in Greek) stills the winds and calms the storm, ensuring that the heroes' continuing travels will be free of interference from wind.

Wind-keeping Aeolus' connection to the element lives on in the aeolian harp, which is one of a handful of human-made instruments played by the wind. Aeolian harps come in many forms, but most involve a wooden box that works as a sounding board, and strings that are stretched across two bridges. The movement of air across the strings creates the harp's music. The harps were placed strategically in the landscape where they might attract a suitable measure of wind to provide harmonious music.

The aeolian harp had a lively reception in Western culture from the eighteenth century, inspiring many poetic and musical works. The role of wind in sounding the aeolian harp has a close connection to the characterization of wind in the *Odyssey* – it is seen as a force connecting the natural and divine worlds. In the Romantic movement, the aeolian harp came to represent humanity's quest for harmony with nature, with harps recognized for their centrality to Romantic thought.[4] The use of the wind harp to give voice to the search for the divine in the natural world is seen in the poem 'The Eolian Harp', written in 1795 by Samuel Coleridge. The poet asks the question, 'And what if all of animated nature,/ Be but organic Harps diversely fram'd ...?'[5] Coleridge described the work as the 'most perfect' of his poems; it concludes by acknowledging the futility of human efforts to comprehend the wonders of the natural and divine spheres. The harp also symbolized the inspiration of the poet – the wind, carrying away old seasons and bringing in new ones, symbolized new ideas and a renewed artistic sensibility, leading to 'an outburst of creative power following a period of imaginative sterility'.[6] The harps reflect the intangible longings of the heart in the works of poets such as Goethe ('Aeolian Harps: A Conversation') and Thoreau ('Rumours from an Aeolian Harp').

Despite their ephemeral quality – or perhaps because of it – winds in classical myth are often tied to wisdom, or the attainment of skill. In the myths of Arcadia, the deity Pan is said to have developed the syrinx, or panpipe, through the inspiration of the wind. The story is recounted in Ovid's *Metamorphoses*, in which the rustic god of the wilderness is described chasing a nymph down to the banks of a river. Trying to catch hold of the nymph, he instead finds himself with an armful of reeds. As Pan sighs in disappointment, a breeze caresses the reeds in his arms and creates a low musical sound. Following the example of the wind, Pan cuts several reeds and creates his eponymous syrinx pipe, bringing it to life with his breath.

Louis Saint-Gaudens, *Piping Pan*, c. 1882, cast 1914, bronze.

Aura on Lucanian
red figure skyphos,
pictured sitting on
a rock by the sea,
430–400 BCE, ceramic.

The relationship between wind and divinity in ancient Greek myth takes numerous forms involving several archetypes and symbols. The Aurai were nymphs of the wind, described by ancient authors such as Aeschylus and Nonnus as swift-winged and wild. The head deity of the Greek pantheon, the sky god Zeus, was often depicted riding a chariot pulled by four winged horses who personified the winds – showing some crossover with the concept of deities riding on the wind in Near Eastern literature. In the writings of the Greek epic poet Quintus Smyrnaeus, the breezy steeds are named as the Anemoi: the four winds. Taken individually, the Greek winds are Boreas the North Wind, Zephryos the West, Notos the South and Euros the East. Known as the Venti in Roman myth, each of the four winds was associated with a season. The invisible Venti are given substantial artistic form in many fountains throughout Rome.

The observation of the creative capacity of wind is reflected in classical myths involving some particularly fertile breezes. In the modern day, the fertilizing capacity of wind is well known: the wind plays a critical role in spreading seeds and pollen so that they may develop into plants. In ancient times, the wind's power in the realm of fertility was widely recognized – but as well as seeding plants, it was thought to act on animals. This theory was presented in several early works in the genre of natural history. Divine, animate winds and horses were features of stories of mares impregnated by winds. The seemingly frequent

Boreas, the
personification of
the North Wind, on
an alabastron from
the Early Corinthian
period, 625–600 BCE.

occurrence of cardinal winds siring mares, either while they grazed in fields or stood on high cliffs, is noted by several ancient authors. Horse-keepers are offered as witnesses to the phenomenon in Aelian's *On Animals* (*c.* second century CE), and Virgil describes the behaviour as inspired by the Roman love goddess, Venus, in the *Georgics* (first century BCE):

> But surely the madness of mares surpasses all. Venus herself inspired their frenzy . . . Love leads them over . . . the roaring Ascanius; they scale mountains, they swim rivers . . . with faces turned to Zephyrus [the West Wind] [they] stand on a high cliff, and drink the gentle breezes. Then oft, without any wedlock, pregnant with the wind (a wondrous tale!) they flee over rocks and crags and lowly dales . . .[7]

Detail of Sprite, depicted by a sculptor/metalsmith close to Donatello, 1432, bronze. The sprite (or 'spritello') combines traits of Mercury and Zephyr, the divine West Wind.

While horses were among the more frequent recipients of fertile winds, other animals too were thought to be impregnated by wind. Classical writers, and medieval authors after them, believed that birds also could become filled with 'wind eggs'. As noted by Aristotle in the *History of Animals*: 'We have cases well authenticated where chickens of the common hen and goose have laid wind-eggs without ever having been subjected to copulation. Wind-eggs are smaller, less palatable, and more liquid than true eggs, and are produced in greater numbers.'[8]

Aristotle describes in detail how birds could take fertilizing breaths and inhale Zephyrus, the west wind. The phenomenon of fertilization-by-wind was further ascribed to sheep by the classical author Aelian in *De natura animalium*. In this case, it is the south wind that is credited with producing the animals' conceptions (rather than the west wind).

While observing the frequent depiction of wind-assisted pregnancies in animals of antiquity, the American historian Conway Zirkle observed that human pregnancies have also, at

Jan Brueghel the Elder and Peter Paul Rubens, *Flora and Zephyr*, *c.* 1618, oil on panel.

times, been considered part of this genre.[9] The view that wind was an active element of semen, causing its foamy texture, had wide currency in the ancient world. The ancient awareness of wind's power of fertilization, extrapolated from plants to animals, likely stands behind the association of winds and spirits in narratives of divine births from virgin mothers.[10]

Away from the genre of natural history, the fertilizing role of wind for flowers and other plants finds mythic representation in the marriage of the Greek deity of the west wind, Zephyrus, and the nymph Flora. In Ovid's *Fasti*, Zephyrus is said to have given Flora power over all flowers as a marital gift.

Wind in Norse and other mythologies

In the world of classical epic, wind shaped stories of heroes facing long stays at sea. In Norse communities, the importance of seafaring made the prominence of winds and storms at the heart of myths and legends a natural fit.

Several major Norse gods can be viewed as wind deities, including Njörd, one of the Vanir gods. The Vanir gods are a small divine group associated with health, wealth, fertility and the ability to see the future. Njörd is the father of two other Vanir deities, Freyja and Freyr, and his primary areas of competence are seafaring, wind and agricultural abundance.

Odin, the chieftain of the Norse deities, and his son, Thor, are closely identified with wind. Adam of Bremen, an eleventh-century German historian, ascribed to Thor the control of 'the wind and showers, the fair weather and the fruits of the earth'. Wind was said to originate with a giant eagle, Hraesvelg, whose name is translated as 'corpse devourer'. Hraesvelg was thought to sit at the world's edge and distribute winds by flapping its wings.

In European myth, the wailing of the wind during inclement weather is personified as a train of ghostly hunters being led on a 'wild hunt' through the sky by howling dogs. 'Ghost riders' are found in numerous cultures; while they are usually connected in some form to wind, the shape of the 'riders' shows some diversity. Many European myths about the wild hunt are considered to have roots in Norse myth. As described by Jacob Grimm, the nineteenth-century philologist and editor of Grimms' fairy tales, 'the phenomenon of howling wind is referred to as Odin's wagon, as that of thunder is to Thor's.'[11]

The natural association of wind with thunder, storms and rain is reflected in the bond between wind and storm deities in many ancient myths. In Norse myth this connection is exemplified by Odin and Thor, and in ancient Persia and India by the close relationship between Vayu and Indra.

Vayu, the Persian wind deity, held a uniting role in the pantheon. He inhabited the realm between the dark earth world of the deity Angra Mainyu, and the light sky world of the prime Zoroastrian deity, Ahura Mazda. Vayu is depicted as a powerful warrior who chases away evil forces. In Indian myth, Vayu is the lord of the winds and breath, and he is frequently accompanied by Indra, the storm god. At times, Vayu functions in the role of the storm deity's charioteer. This divine quality reflects the natural capacity of winds to carry along other weather events observed in the natural world. Vayu is sometimes called *Sadagata* ('ever-moving') or *Gandhavaha* ('bearer of perfumes'), reflecting the potential for the invisible force of wind to be perceived through other senses, such as smell and touch.[12]

Wind in Mesomerican, Maya and Aztec myth

Maya myths include winds of all kinds, with different varieties of wind representing different deities. Huracan is the eldest thunderbolt god in Maya myth, and his name means 'one-footed' – referring to the single point at which a hurricane connects to the earth. Huracan is sometimes called the 'Heart of the Sky', and, as with wind's natural role, his myths involve creation and destruction – notably through a great flood that destroyed humanity in the distant past, and in the creation of life that followed.

In 2015 archaeologists working in Belize unearthed an artefact covered in Maya hieroglyphs.[13] This has been called the 'wind jewel' and it is the only one of its kind to have been discovered

Figural pendant made of jadeite, Maya, 450–600 CE.

bearing a written text. The jewel itself takes the form of a jade pendant, worn around the neck of the king during ceremonies. Along with the jewel, the archaeologists found a vessel with a beaked face, likely depicting the Maya wind deity. The wind jewel is fashioned in a long 'T' shape and is also carved with a 'T' on one side. This sign is the glyph T503 IK', meaning 'wind' and also 'breath'.

Interestingly, earlier varieties of 'wind jewels' discovered by archaeologists often contained fragments of conch shells. These ornaments were worn by priests of the feathered serpent deities Quetzalcoatl and Ehecatl: depictions of Quetzalcoatl show that he wore a wind-jewel upon his pectoral. Queztacoatl and Ehecatl are pre-Columbian deities, both with ties to wind. Often the two deities are conflated, with Ehecatl considered an aspect of Quetzalcoatl. In Aztec myth, Ehecatl uses wind to blow on the Sun, setting it in motion on its journey through the sky, after the gods create the Sun and Moon. Further 'wind' from human work, prayer and sacrifice is thought to keep the Sun in motion. The connection of the conch shell to the wind jewel on Quetzalcoatl's breast is thought to represent the belief that when holding a conch to one's ear, the rushing sound inside is made by the blowing of the wind (much as today we might think of the sound of the ocean when listening to a conch shell).

The recent discovery of the jade wind jewel in Belize shows a thematic connection between the jewel's material and its meaning. Jade was a ubiquitous feature of Classical Maya art, inextricably linked to concepts of rulership and centrality.[14] At the same time, jade represented wind and breath. The ability of

jade to feel cool to the touch, but warm once held in the hand, contributed to the Maya view that jade was a breathing, animate substance and an incarnation of living breath.[15] Wind's vital role in carrying rain was linked in Maya thought to its identification with the essence of life. This connection is further seen in the conception of the Maya deity Chaac, who was the lord of thunder, lightning, wind, rain and fertility. Similarly, the connection between jade, wind, breath and spirit stretches far into the distant past. The analysis of burial sites from the Formative period (1000 BCE–500 CE) finds pieces of jade placed in the mouth of the deceased, likely to enable receipt of the breath and life-force of the dead. This use of the jade 'breath bead' also occurs in Mesoamerican sites from the Middle Formative Olmec to the Late Postclassic period.[16]

Quetzalcoatl Teotihuacan Plaza.

Thirty hieroglyphs on the back of the wind jewel appear to describe a religious ceremony (although the Maya script is not fully deciphered, nor the subject of scholarly consensus). The ceremony described on the back of the pendant involves 'scattering', which likely refers to the known ritual scattering of incense by religious specialists (including royalty) to bring on the wind and rain. The magical properties of the wind jewel made it a critical part of scattering ceremonies, and it can be seen being worn on the chest of a central participant in iconographic evidence related to the find. The jade wind jewel was first used for this purpose in 672 CE, when it belonged to a king named Janaab' Ohl K'inich. The jewel was buried on a T-shaped platform many years later in

Market-woman figure, wearing conch-shell-shaped earrings representing the 'wind jewel', Mexican (Nopiloa), 600–900 CE, earthenware.

850 CE. The archaeologists who made the discovery note that the timing of the wind jewel's burial corresponds with the collapse of the Maya civilization, a result of climate change.[17]

Wind in African mythology

The power of wind to bring rain is also a feature of its role in West African myth. Shamans and rainmakers summon winds when the land needs rain, and in Liberia the practice is linked to the story of the god Meleka and his daughter Sia. The suffering of humanity caused Sia to cry, flooding the land, and Meleka called on the winds to distribute the divine deluge evenly.[18]

The connection of wind to rain and storms is found in the mythology of the Bushmen of the Kalahari Desert. The creator deity Kaang is described as a god of wind, rain and breath. He is seen as a powerful yet invisible spirit who gave life to all living things, as well as to natural phenomena such as the Moon, thunder and the wind. The universal origins of the spirits within living things meant spirits were thought capable of moving between different forms – for example, the spirit of a man may fly into the body of a lion in Bushmen myth. When a person dies, the spirit or wind within them is thought to arise from their body and blow away their footsteps, removing every trace of them from the earth.

In the Zulu cosmic worldview, the wind is used in divination to bring messages from the spirit world, and it is closely tied to witchcraft. In Zulu cosmology, humans consist of three main parts: the body, the shadow and the *umoya* – a word referring to spirit, soul, air, wind and breath.[19] The close connection of wind and the spirit in Zulu traditions reflects a long-standing relationship between the spiritual and weather worlds, reflected in the shamanic practice of wind manipulation.[20]

Wind, breath and air

The origins of breath too are frequently linked to wind – Navajo philosophers understood wind, or *nilch'i*, as the breath that animates all life, and it is thought that this wind takes individual shape in each creature: 'It is said that the wind that animates life is visible in the whorls in one's fingerprints and in the dust storm funnels ("ghost riders") that twist and gallop across open vistas.'[21] This view on the nature of breath and its distinctive shapes, found in Navajo philosophy, shows intriguing parallels with recent scientific research on the subject. A 2013 study by researchers in Zurich found that each individual's exhalations hold a unique molecular signature – like a 'breath fingerprint'. These 'breathprints' can be used to assist in medical diagnoses and have the potential for authenticating identity.

Moving briefly further east, the coupling of wind energy and breath famously are a well-established feature of Buddhism, as seen in the works of Rangjung Dorjé, the third Karmapa of Tibetan Buddhism. In the thirteenth century, Dorjé became the teacher of Toghon Temür, the eleventh emperor of the Yuan Dynasty, and the fifteenth Khan of the Mongol Empire. On the night that Dorjé died in the year 1339, it was thought that his face appeared in the full moon, a vision witnessed by Dragpa Sengge, a Mongol chieftain. In his description of the tantric cosmos in the short treatise 'Key to the Essential Points of Wind and Mind', Dorjé describes the wind (which can also be translated as 'breath' or 'energy') as mediating between the mind, body and world, stating that 'in short, wind is thought.'[22]

Returning to First Nations myth, wind features prominently in the great Navajo creation myth 'Diné Bahane' (Story of the People). The story begins with the Holy Wind (*Niłch'i Diyin*) being created in mists of light out of darkness and bringing the four Holy People to life. Later, First Man and First Woman are created from ears of corn, and life is breathed into them by the wind:

It was the wind that gave them life. It is the wind that comes out of our mouths now that gives us life. When this ceases to blow we die. In the skin at the tips of our fingers we see the trail of the wind; it shows us where the wind blew when our ancestors were created.[23]

As well as being a creative force in Navajo belief, the Holy Wind binds everything in the cosmos together, and joins the inner and outer worlds. In this way, breathing is understood as a sacred act, animated by the force that unites all beings and the world in which they exist in a dynamic union.[24]

The concept of wind as a divine connecting force is also found in Iroquois myth, and featured in the influential ethnological work *The League of the Iroquois*, first published by the American anthropologist Lewis Henry Morgan in 1851. As described by Morgan, Gaoh was a giant deity who personified wind and was thought to live in the far northern sky. Gaoh's role involves controlling the four winds and keeping them in balance, with each cardinal wind named for a different animal: the north wind (Bear), east wind (Moose), south wind (Fawn) and west wind (Panther).[25] It was Gaoh who in Iroquois legend named each of the four winds and gave them their roles and their dwelling places.

In the myths of the Lakota Sioux, the wind god Tate is a benevolent spirit who is formed as a companion to the sky, Skan. Tate has no physical body but appears in a shadow-like form. As with the Iroquois myth, the wind deity of the Lakota Sioux is related to the four cardinal winds. The cardinal winds are Tate's sons, and each takes a turn every year to bring forth his own season. Eya, the deity of the west wind, brings rain and thunder; Yanpa is connected to the east wind and thawing winds; Waziya is the god of the north wind, who brings ice, frost and snow; and Okago of the south wind brings warmth and fair weather.[26] The four winds were brought to earth by Tate to accompany his beloved partner, Ite. While Tate remains largely apart from the human world, it is through his sons, the four winds, that he connects with humans and influences their lives.[27]

The winds of Polynesia

The northeast trade winds have played a shaping role in the climate and culture of Hawaii. As well as providing energy for sailing and the spread of fertilization, winds provide information to humans through divination. The Hawaiians have weather experts who specialize in reading the winds, called the *kahuna kilo makani* (observer of the winds), who, like the experts in reading the stars and Moon, can interpret the meanings of signs found in the natural world.[28] The wind deity Paka'a is credited with inventing the sail in Hawaiian myth. The story of Paka'a comes from a relatively late period in the history of Hawaii's ruling chiefs, but is thought to build upon older traditions from the islands.[29] The story sees Paka'a inheriting his power over winds from his grandmother Loa, who had controlled the wind across every district of the islands from Hawaii in the east to Kaula in the west. Using his grandmother's wind gourd, Paka'a summons the wind of his choosing by calling its name. The invention of the sail by Paka'a ties him to Hawaiian wayfinding traditions, as does his identification as a navigator who could read and interpret the signs of the sky, stars and wind.

New Zealand myth

In Māori myths the four cardinal winds are also conceived of as the children of the wind deity. Tawhirimatea is the god of wind and storms in Māori tradition, and the son of Papatūānuku (earth mother) and Ranginui (sky father). With his wife, Paraweranui, Tawhirimatea had numerous offspring, known as Wind Children.

Tawhirimatea takes a stand against his birth family in the Māori creation myth that details the origins of the arts of agriculture, fishing, hunting and cooking. In one version of the story, the continual union of the sky father and earth mother in Māori myth causes friction among their powerful children. The siblings agree to try and separate their parents, except for Tawhirimatea, who disagrees with his siblings. He sends his wind children away, each to a different cardinal point: tūāraki to the north, tonga to the south, marangai to the east and hauāuru to the west. Tawhirimatea's siblings are eventually successful in separating the earth mother and sky father, but a divine battle results, causing continuing friction between the wind deity and his family.

Māori traditions relating to weather involve highly skilled religious specialists called *tohunga*, who are able to 'raise the wind' using powerful incantations. The winds can be used to bring favourable weather conditions for their communities, or to create bad weather for an enemy.

From before the time of the first humans, wind has shaped the earth. Across diverse civilizations and cultures, it has exerted a similarly potent formative influence on the stories, myths and legends that provide structure for human conceptions of the natural and supernatural worlds.

Winds in myth and religion are frequently shown to be powerful forces which may be creative or destructive. The role of wind in many myths and legends goes beyond that of a primary character shaping timeless narratives: wind has also played an important role in spreading human stories and ideas across the globe. As with the spreading of pollens and seeds, wind lifted the sails that allowed for the exchange of ideas and trade across cultures in the ancient world. This germinating role is reflected back in the stories themselves, in which winds are a familiar fixture in stories of seafaring, distant lands and ancient origins.

4 Warring Winds

The wind's capacity to enact change is one of its most universally recognized features. Changes brought about by wind can be slow and imperceptible, but at times raging winds and sudden squalls have dynamically blown their way into the history books. This chapter considers the historical significance of warring winds, both as powerful historical wind storms and as forces shaping the course of human conflict.

The destructive force of damaging winds has inspired human efforts to quantify and counteract their effects. The invention of the Beaufort scale, by which winds are measured, was inspired by the 'Great Storm' that struck Britain in 1703. The storm was considered the most powerful to have occurred in Britain's recorded history, with the resulting coverage in newspapers described as the 'first media weather event of the modern age'.[1] Daniel Defoe, the English author of *Robinson Crusoe*, was inspired by the storm to propose a 'Table of Degrees' for describing wind forces – a precursor to the Beaufort Scale.

Wind has played a decisive role in numerous battles that have shaped the course of history, as witness the *kamikaze* ('divine wind') that were credited with twice saving Japan from the invading Mongol fleets of Kublai Khan. In modern times, the role of wind in warfare has amplified the potential destructiveness of human conflict. The dispersal of poisonous gases and toxic substances has increased the ability of war to extend its reach far beyond the field of battle, carried on the wind.

Diagram of the winds, from Isidore of Seville, *De ventis*, in a cosmographical manuscript, late 12th century.

Weaponizing wind

From the time of our earliest written records, wind and warfare have been intertwined. The appearance of wind in warfare in ancient times was connected closely to divine power and favour. From ancient Mesopotamia to the modern day, wind assistance in battle has proved one of the most critical – yet least predictable – elements of warfare.

In the modern age, innumerable historical conflicts have seen wind harnessed in the service of making war. Wind has filled the sails of warships, buoyed fighter planes, carried lethal gases and scattered devastating radiation. Yet despite wind's immense power, its historical use in battle has proved a delicate exercise. The favourable wind of myth, with its promise of swift victory, has proved a fickle and even dangerous ally in the real world. The connection between wind and change is tangibly clear on the battlefield, where a sudden breeze, change of wind direction or even an unexpected tornado can rapidly alter the outcome of battle – and the course of history with it.

Along with battles themselves, wind has played a pivotal role in preparations for war throughout history. In 1954 the u.s. military's testing of nuclear fusion bombs on Bikini Atoll was adversely affected by an unexpected weather event. The strength of the bomb detonated in the Castle Bravo test on 1 March proved to be more than double its predicted size – and over 1,000 times the size of the nuclear bombs dropped on Hiroshima and Nagasaki in the Second World War. The wind on 1 March in Bikini did not follow the patterns predicted by the meteorologists. As a result, strong westerly winds carried fallout contamination across the population of the Marshall Islands, and beyond. More than seventy years later, Bikini Islanders continue to face the consequences of the spread of radiation from the nuclear tests.[2] Wind shear and ocean currents carried the fallout as far as Europe, Australia and India, causing an international outcry. Although the tests proved how vulnerable the new weapons could be to a change of weather, and the lethal consequences that could ensue, in the following four years another 21 nuclear bombs were tested at Bikini.

Wind in battle

In recent years, awareness has been building of the contribution of weather patterns to historical human events. From the Roman Climate Optimum, in which warm, stable weather conditions coincided with a period of political balance in the Roman Empire, to the wind-driven waves that threatened to delay the largest seaborne invasion in history at Normandy in 1944, weather has played a decisive yet overlooked role in significant historical events. Weather and warfare share the unpredictable capacity to create destruction, upend the status quo and change the course

'Baker Day' atomic bomb test, Bikini Atoll, Micronesia, 25 July 1946.

of history. When weather and warfare combine, the results can be as decisive as they are unforeseeable.

Hannibal

The libeccio is a warm westerly or southwesterly wind that blows through parts of the coastal Mediterranean. It is known by several names and is notorious for bringing high seas. Recently, the libeccio wind has lent its name to an international anti-drug-trafficking collective in the waters around Italy and Corsica, but it is arguably most famous for its role in the ancient Battle of Cannae between the Romans and the Carthaginians.

In 216 BCE, Hannibal led the Carthaginians to victory over the larger Roman army in a bloody battle near the ancient village of Cannae in southeastern Italy. In previous battles against the Romans, Hannibal had demonstrated a talent for surprising his opponents by using the natural features of the terrain to aid his attacks. In the Battle of Lake Trasimene, he hid his troops in a valley covered by mist to enact one of the most famous ambushes in military history. At Cannae, he used the wind and the Sun to defeat his adversaries.

Hannibal's skill at reading terrain meant he was aware that the direction of the libeccio wind could prove a decisive element in the battle. Knowing that the wind would intensify in the heat of the afternoon, he positioned his troops so that the wind would blow against their backs, and into the faces of his enemies. This strategy was recorded by the Roman historian Livy:

> It was in the neighbourhood of this village that Hannibal had fixed his camp with his back to the Sirocco which blows from Mount Vultur and fills the arid plains with clouds of dust. This arrangement was a very convenient one for his camp, and it proved to be extremely advantageous afterwards, when he was forming his order of battle, for his own men, with the wind behind them, blowing only on their backs, would fight with an enemy who was blinded by volumes of dust.[3]

The name given to the wind in Livy's account, Volturnus, is also the name of a river and a deity. It is thought to come from the Latin word *volvere*, meaning 'to roll along' or 'to wind around' – the same Latin word lends its name to the Volvo car brand. The Volturus wind was considered to have dangerous and destructive qualities: the Roman poet Lucretius described thundering on the heights, while the Roman writer Columella observed how the wind scorched the grapes on the vine with its fiery breath.[4]

The wind blew dust and grit into the eyes of the Roman fighters, creating a distracting irritant and impairing their vision. The importance of unimpeded sight on the battlefield would have been well understood by the one-eyed Carthaginian leader: Hannibal himself had recently suffered an eye injury on campaign; he wore an eyepatch after the Battle of the Trebia in 218 BCE, probably due to the loss of an eye. While most historians agree one of Hannibal's eyes was lost to an infection, some suggest he had ophthalmia, a chronic inflammation of the eye. In spite of this, Hannibal's ability to read the wind handed him a momentous victory over the mighty Roman army.

Kamikaze: divine wind

The ancient association between wind and divine favour recurs in the more recent martial history of Japan. Famously, two great winds saved the Japanese nation from invasion by the Mongols in the thirteenth century.

At its height, the Mongol Empire was the largest contiguous land empire in history. In 1274 CE, an attack by the Mongol fleet proved successful in conquering the Japanese islands of Tsushima and Iki. The invaders were repelled by samurai at Hakata Bay, however, and as they began to retreat, there was a sudden change of weather. The enormous Mongol fleet that was beginning to withdraw from Hakata Bay was savagely battered by the arrival of a sudden typhoon. The typhoon's arrival devastated the fleet, resulting in the loss of some 13,000 men and a third of the Mongol ships. The wind event was considered 'highly fortuitous'

Kikuchi Yōsai, *Mongol Invasion*, 1847, watercolour depicting the destruction of the Mongol fleet in a typhoon.

by the Japanese; the awareness of their miraculous escape from invasion led to the building of defensive walls along the coast to prepare for their enemy's return.[5]

Seven years after the defeat at Hakata Bay, the Mongols joined with their Chinese and Korean allies and sent a larger fleet to invade Japan. Initially rebuffed by walls built by the Japanese in the intervening years, the Mongol fleet searched for a safe entry point. The Mongol forces greatly outnumbered the Japanese fighters, but once again the weather turned against them. A second typhoon destroyed more than half of the Mongol fleet, and many of the survivors were killed by the waiting Japanese. The Mongols made no further attempts to invade Japan, after having twice been devastated by the weather events known by the Japanese as the *kamikaze*, or 'divine winds'.

The legendary status of the *kamikaze* winds has raised questions over their historicity. Detailed archaeological explorations of the waters around the Japanese islands have found large numbers of sunken shipwrecks that have been attributed to the battle.[6] Further support for the historicity of the 'divine winds' has been found from the analysis of pulverized shells and changes in sediment in the region, consistent with damage from a giant typhoon.[7] Yet the long passage of time since the rule of Kublai Khan and the appearance of the invading Mongol fleet has raised uncertainty over the significance of the role of the 'divine winds' in the two Japanese victories. Recently, greater emphasis has been placed on the importance of the robust Japanese resistance, led by Hojo Tokiume, the eighth shikken of the Kamakura shogunate.

The importance of soldiers and monks in the defence of the Japanese nation saw their social status increase during the Kamakura period (1185–1333), compared to the Heian period (794–1185) with its privileging of courtiers. Thus the successful defence against Kublai Khan to which the 'divine wind' contributed brought a change that carried far beyond the battlefield. The *kamikaze* winds show how a change in the weather can effect substantial social and political change, and the union of religion and warfare in human thought. The two typhoons are testimony to the way in which 'increases in severe weather associated with changing climate have had significant geopolitical impacts'.[8]

The War of the Roses

Historically, archery has been shown to be one of the battlefield components most influenced by the wind. A tailwind boosts the archers' range, allowing them to fire on the enemy effectively from a safer distance. Conversely, a headwind limits the arrow's range and may interfere with its intended flight path. The effect of wind on archery proved to be a 'crucial' factor in the outcome of the famous Battle of Towton, fought between the Yorkists and the Lancastrians during the English War of the Roses (1455–87).[9]

The Battle of Towton has been described as the largest and bloodiest battle to take place on English soil. On 29 March 1461, some 50,000 soldiers fought on Palm Sunday in a snowstorm, the battle lasting four hours. The troops of the Lancastrian king, Henry vi, had the positional advantage: they occupied the high ground on the battlefield, between the North Yorkshire villages of Towton and Saxon. Yet this proved to be of lesser importance than the weather conditions. The Yorkists, led by the eighteen-year-old Edward, had the wind at their backs. The strategic positioning of troops to take advantage of the wind conditions has been credited to Lord Fauconberg (William Neville), a prominent Yorkist. At the age of sixty, 'Little Fauconberg' had a lifetime of experience on the battlefield and was adept at reading the weather.[10] The Yorkist arrows pierced deep into the body of the Lancastrian side. The Lancastrian arrows, flying into a powerful headwind, failed to reach the enemy's position. At the same time, the wind drove sleet from the storm directly into the eyes of the Lancastrian archers, inhibiting their aim.

The range and accuracy of the Yorkist archers, who had been able to 'measure the wind', forced the Lancastrians to surge forward, shortening the distance between the two forces and conceding the high ground.[11] Heavy fighting commenced, with the tide turning in favour of the Yorkists after the arrival of a company led by the Duke of Norfolk. The eventual Yorkist victory saw Henry VI flee to Scotland with his wife and son, and greatly reduced the power of the House of Lancaster in England.

The Spanish Armada

In May 1588, a fleet of 130 ships set sail from the Spanish city of La Coruña, destined to take part in an invasion of England. The ships were under the direction of their newly appointed commander, the Duke of Medina Sidonia, Alonso Pérez de Guzmán y Sotomayor. The Duke's last-minute appointment to the role followed the abrupt demise of the fleet's organizer, the noted Spanish admiral the Marquis of Santa Cruz. It was Santa Cruz who first called for the Spanish Armada's attack on England in a letter to the Spanish king, Philip II, in 1583. Famously undefeated in battle throughout a remarkable military career spanning fifty years, the Marquis identified England as a threat to the Spanish empire and advocated for war. Philip put Santa Cruz in charge of preparing the fleet.

In contrast to his predecessor, the Duke of Medina Sidonia was not experienced in commanding a naval fleet. He petitioned the king to reconsider his appointment (citing his own seasickness among numerous reasons for a change of mind), nor did his lack of experience prevent him from foreseeing the problematic path ahead for the Spanish Armada.[12] What no one could predict, however, was the unprecedented role the winds at sea would play in turning the fortunes of the battle in England's favour.

The passage of the Spanish Armada towards the shores of England was delayed by bad weather, including high winds, dangerous seas and storms. The Armada was intended to rendezvous with a huge army that had been assembled in the Spanish Netherlands. The fleet would be used to shepherd the army in unarmed barges across the English Channel, where they would make landfall close to London. A combination of unfavourable weather and fortuitous strategies from the English fleet meant that the troops were left waiting for a transport that would never arrive.

At the end of July, the Spanish Armada attempted again to join up with their waiting land troops, but strong south-westerly winds thwarted their efforts. After a few indecisive encounters with the British, the Armada was ordered to sail north in mid-August. Following these orders, the Spanish fleet found itself beset with stormy weather for the rest of the voyage. Hurricane-force winds drove many of the Spanish ships onto the Irish and Scottish coasts, where they were wrecked. Over fifty vessels were lost, and the remains of the fleet turned for home.

The failed conquest by the Spanish Armada was received in England as a victory of Protestantism over Catholicism – continuing the ancient theme of linking weather, warfare and divine favour. The weather that sank the Spanish ships became known as the 'Protestant wind', and the divine origin of the weather was emphasized in medals celebrating the English victory. Several medals carried the inscription 'Jehovah blew with His winds, and they were scattered.'

Armada medal, 1588, commemorating the defeat of the Invincible Armada and bearing the inscription *Flavit Jehovah et dissipati sunt* ('Jehovah blew and they were scattered').

The American Revolution

During the American War of Independence (1775–83), the Battle of Long Island (also known as the Battle of Brooklyn) was the first major battle following the Declaration of Independence by the United States on 4 July 1776. In terms of troop numbers deployed, it is generally considered the largest battle of the entire seven-year war.

The Battle of Long Island took place in the modern-day borough of Brooklyn, along the western edge of Long Island. Weather had a decisive impact on the outcome of the conflict, and the shape of the war to come. On 22 August 1776, George Washington and his troops in Brooklyn confronted an advance guard of British troops, who had arrived from Staten Island. Washington had been moving troops into the region for several months, as he rightly believed the area's strategic significance would make it a likely target for the British to attack. The main conflict began on 24 August and ended in a solid victory for the British army and Hessian troops. A final series of assaults around Vechte-Cortelyou House (now known as Old Stone House) found Washington and his army surrounded by enemy troops, pinned down in Brooklyn Heights with the East River at their backs. The British laid siege to Washington's position, but the prevailing weather conditions worked against them. On land, their advance was slowed by rain, and the direction of the wind prevented the English commander, Admiral Howe, from sending ships up the river to attack the trapped troops in a pincer movement. Washington knew that this line of attack from Howe would arise as soon as the Admiral had 'a favourable wind'.[13]

After losing a fifth of his fighting force and outnumbered two to one, Washington finally decided to retreat. He sent a letter to General William Heath, stationed nearby, to send every boat available so the troops could make their escape across the river. The need to keep the patriot's night-time departure secret meant that silence was crucial. The wheels of wagons and oars of boats were muffled, and the retreat was further concealed from their adversaries by the best possible weather conditions, including a 'miraculous' shift in the wind for the patriots' sailboats and the appearance of a heavy fog.

Washington's fortunes were further bolstered by the elements in subsequent battles. The movement of the wind also halted the occupation of the city of Washington by the British in the War of 1812. The arrival of a tornado soon after the attacks on the city by British troops sent the invading forces into retreat. The event became known as 'the storm that saved Washington'.[14]

Winds and gas in warfare

The use of poisonous gas deployed on prevailing winds in the First World War is often noted as the first large-scale use of a weapon of mass destruction (chemical, biological or nuclear).[15] The development of gas warfare utilizing chlorine, mustard and other gases inflicted psychological horror alongside devastating casualties and lifelong injuries. Wind and other variables, such as temperature and precipitation levels, were carefully measured by military meteorologists, who advised on the optimal time to release the gas to cause the greatest damage to the enemy forces.

Yet wind in warfare, as we have seen, is an unpredictable and often dangerous ally. A change in the wind's direction or a shift in its intensity could result in unintended consequences – and, potentially, blowback. The nebulous quality of gas borne on the wind meant that the poisons could not be restricted to the battlefield and could easily carry on the breeze to villages near the battlefront, causing civilian casualties. Despite alarm around the destructive potential of the new gas weapons, and several treaties to ban their use, chlorine gas was deployed by German troops in 1915 during trench warfare in Belgium. It is thought that the use of gas in the First World War caused more than a million deaths. Following the First World War, the dispersal method went through many changes, such as moves to chemical shelling or missiles, likely due to the difficulties and dangers involved in the use of wind.

The capacity of wind to carry dangerous substances long distances has had unexpected consequences. A French study

from 2018 has shown that wines from Napa Valley in California exhibited a higher level of the isotope caesium-137 following the Fukushima disaster of 2011.[16] The authors of the study were interested in discovering whether the Fukushima disaster might have a similar dispersal of radiation through the air by wind to that found following the accident at the Chernobyl nuclear power plant in 1986.

Whey in the wind.

Destructive wind and weather events

Winds and warfare are two areas in which abrupt and capricious change can upend the established order. The combination of power and unpredictability in both areas has historically seen weather and warfare closely aligned with supernatural forces, in the awareness that controlling the path of weather and war lies (mostly) beyond the limits of the human condition. The second part of this chapter explores major weather events involving wind, and how catastrophic weather events illuminate the realities of the human condition and our connection to the environment in which we live. As Jonathan Bate has observed in his analysis of the 'weather poems' of Byron and Keats, 'The weather is the primary sign of the inextricability of culture and nature.'[17]

In contrast to the often-orderly march of the seasons, the gradual procession of the planets and the regularities of predictable weather patterns, major weather events provide bracing evidence of the primal, potent natural forces that constitute the world around us. The natural forces that drive unexpected and catastrophic weather events, such as volcanoes and violent winds, are the same forces that moulded Earth into its now-familiar form. Catastrophic weather events show how these primal forces are still a part of the world around us, shaping our past, present and future. Historian Jan Golinski well encapsulated the connection between humans, weather and the course of civilization:

> The regular climate was viewed as benevolent, a force that integrated human beings with their environment, that answered to the needs, and could even be modified by the progress of civilization. But occasional weather disasters brought out profound doubts about progress, about its limits and its drawbacks . . . The weather apparently still has the power not just to disrupt our material lives, but also to make us reflect on the shallowness of civilization.[18]

The arrival of damaging winds and impactful storms draws focus on the often-invisible work of weather. Powerful wind events can result in destruction and chaos, but also provide a sense of awe at the power of nature and the resilience of human life.

Hurricanes, cyclones and typhoons

Wind regularly performs as an unseen, creative force in nature. When it comes to catastrophic weather events, however, wind can be at its most dangerous when it is most visible. Examples of these events involving powerful winds are cyclones, typhoons and hurricanes. Typhoons, hurricanes and cyclones are rotating storm systems with areas of low atmospheric pressure at their centres, and all involve strong winds, rain and spiralling movements. Differences in the definitions of storm systems are based on the place where they develop. Storms that develop in the Atlantic Ocean and northeastern Pacific Ocean are hurricanes. In the northwestern Pacific Ocean, they are called typhoons, and in the South Pacific or the Indian Ocean these storms are tropical cyclones. These terms show some interchangeability in different contexts and in different parts of the world.

In a cyclone, the system of winds in a low pressure area will spiral inwards. Because of this cyclical motion, cyclones take their name from the Greek word for 'circle', *kuklos*. A hurricane is a storm with cyclonic winds that measure 119 km/h (74 mph) or more. It is thought that the word 'hurricane' came via the Spanish *huracán* in the mid-sixteenth century. This Spanish word itself derives from *Jurakan*, a Taino word for 'god of the storm'. As well as giving hurricanes their name, Taino people held a sophisticated insight into the movements of the storms, hundreds of years before these were rediscovered with modern technology during the Second World War. The Taino were an indigenous population of the Caribbean, and the first New World peoples to be encountered by Christopher Columbus in his voyage to the Americas of 1492. The early Spanish explorers were told of a powerful wind deity, who signalled divine anger through the movement of winds, with Columbus hearing the local people speaking of an unimaginable natural power called Hurakan.[19]

Hurakan (or Jurakan) is thought to be related to the Maya deity Huracan, a creator god connected to wind, fire and storms. It is uncertain whether Jurakan of the Taino people was originally considered a male or female deity, as depictions give no clear sign of the figure's sex. Jurakan is depicted with a head, no torso and two spiralling arms. The majority of these images contain cyclonic, or counterclockwise, spirals, likely inspired by the tropical cyclones that were native to the Caribbean.[20] This artistic feature shows

Tropical Cyclone Chapala as the eye of the storm was approaching Yemen's coast, 2015.

intriguing alignment with natural features of cyclonic motion. It is thought that the Taino people 'discovered the vortical nature of hurricanes several hundred years before the descendants of European settlers did'.[21]

Hurricane moving across an open field.

The spiralling patterns now associated with cyclonic winds were only discovered in the modern day when meteorological radar was developed during the Second World War, as military radar operators observed the echoes of weather events while monitoring enemy targets. As the Caribbean is in the northern hemisphere, the spiralling winds of cyclones there would move in a counterclockwise (or 'cyclonic') direction. Due to the Coriolis effect, cyclones of the southern hemisphere spiral in a clockwise ('anti-cyclonic') direction. The Coriolis effect is the apparent deflection of objects (including missiles and storm winds) that are moving in a straight path across Earth's surface, due to the planet's rotation. It is named after the French engineering professor Gustave-Gaspard de Coriolis, who first discovered it in 1835. The effect becomes more powerful further away from the equator, towards the poles. The Coriolis effect influences ocean currents as well as wind, as the currents are driven by the movement of air across Earth's surface.

Keeping records of the wind

The capacity of strong winds to create destruction and reshape terrain has made important work of not only cataloguing and quantifying the movements of storms but attempting to introduce some order into the chaos they bring. With the success or failure of long and dangerous sea voyages often predicated on the favour of the wind, the logbooks of seventeenth-century mariners reveal keen observations of weather events – and an awareness of the need for a common vocabulary in describing winds.[22] Among many pioneers of meteorological measuring was the English colonist Captain John Smith – most famous for his connection to Pocahontas – whose book *A Sea Grammar* (1627) included a chapter on wind terminology.[23] At the turn of the eighteenth century, a catastrophic weather event struck England which would change the way that weather would be communicated, and advance the cause of finding a universal language for wind.

In 1703 the 'Great Storm' hit central and southern England, bringing unprecedented destruction and hurricane-force winds. Thousands of people were killed, ships were wrecked, and buildings and trees were felled. The storm winds were sufficiently strong to strip Westminster Abbey of its lead roofing. In his book *The Storm* (1704), the English writer Daniel Defoe catalogued the damage. His account of the storm is credited as the 'first weather media event' of the modern age, owing to the popularity of his account and its influence in shaping the memory of the event.[24] Defoe's *Storm* included a 'Table of Degrees', a chart that quantified different types of wind on a scale starting with 'stark calm' and ending with a tempest.[25] Defoe's effort to categorize different types of wind into a scale of increasing severity shows commonalities with the famous wind scale developed over a century later, and which is still in use today – the Beaufort wind force scale.

In the logbook of an Irish hydrographer in 1806, the Beaufort scale was created. Sir Francis Beaufort had already had an eventful naval career before making his legendary contribution to meteorological science with the scale that bears his name. An early sea voyage on a ship owned by the East India Company ended in shipwreck, with the vessel's treasure chests tossed into the sea, and he was involved in several entanglements with pirates. Moving from the merchant service to the Royal Navy, Beaufort participated in several sea battles, surviving a near drowning, some sabre wounds and 'a point-blank volley in the face from a blunderbuss' while leading the boarding of a Spanish ship.[26]

In his mariner's logbook in January 1806, Beaufort made an entry that would significantly shape the future quantification of weather: 'Hereafter I shall estimate the force of the wind according to the following scale'. A list followed, numbering thirteen different types of wind from 'calm' to 'storm'. Although much refined in the two hundred years that followed, the Beaufort scale was adopted as standard by the Royal Navy in 1838, and its use then spread to non-naval vessels. At the first meeting of the International Meteorological Society in Brussels in 1853, Beaufort's scale was accepted as generally applicable. By using the scale, mariners were able to estimate the force of prevailing winds through visual cues, and communicate what they saw with a measure of standardization.

With the growth of steam power, and the decline of sail, Beaufort's weather markers were updated by George Simpson, the director of the United Kingdom's Meteorological Office, to

Warring Winds

Beaufort's diary from 1806 showing his original scale.

include geographical descriptors. Although a great leap forwards in weather science, the process of developing and refining the Beaufort scale reflects the continuing challenge of quantifying the wind – a complex and often-imprecise exercise, even in the modern day.

Efforts at creating a universal scale or language for measuring winds and the powerful force of storms has allowed for the collection of records and the ability to compare the impact of diverse weather events. On 12 October 1979, Typhoon Tip showed peak wind speeds of over 300 km/h (185 mph), making it the largest cyclone ever recorded. As well as its ferocious winds, Tip is known for its massive size: making landfall in southern Japan, its area of circulation spread across 2,200 km (1,360 mph), making it the largest storm on Earth.[27] Typhoon Tip caused widespread destruction, with the loss of more than 20,000 homes and the deaths of 99 people.

Typhoon Tip is considered a 'supertyphoon'. Cyclonic storms are categorized as 'super' when their wind speeds reach more than 240 km/h (150 mph). The most powerful winds ever recorded in a supertyphoon are attributed to Typhoon Goni, which made landfall in the Philippines on 28 October 2020. It was the strongest landfalling cyclone in recorded history, with winds exceeding 315 km/h (195 mph). As well as the devastating loss of lives and property, Typhoon Goni is thought to have caused u.s.$415 million in damage.

The intensity of hurricanes is measured by wind speed on the Saffir–Simpson Hurricane Wind Scale (sshws). Storms are ranked from categories one to five, with category five being the highest. The sshws was developed in 1969 by Herbert Saffir, who was working for the United Nations studying low-cost housing in hurricane-prone regions. Inspired by the Richter magnitude scale for earthquakes, Saffir created a system of classification based on wind speeds, to which meteorologist Robert Simpson added storm surge and flooding. The record for the most powerful sustained winds in an Atlantic hurricane is held by Hurricane Allen, in August 1980. Hurricane Allen hit the Caribbean, Mexico and southern Texas, and also holds the record for the hurricane to have spent the longest time at Category 5 intensity.

While wind speed is a standard measure of a hurricane's impact, there are many more factors to consider. In terms of human cost, the deadliest hurricane on record is the Great Hurricane of October 1780, which killed more than 22,000 people as it passed through the Lesser Antilles in the Caribbean Sea. The hurricane with the most intense minimum barometric pressure was Hurricane Wilma. A hurricane's barometric pressure is a key indicator of its intensity, with lower pressure reflecting greater intensity. Hurricane Wilma was one of several Category 5 hurricanes in the record-breaking 2005 season – a season that has only been surpassed by 2020 for the severity of Atlantic hurricanes.

In terms of measuring financial impacts of cyclonic weather events, the costliest hurricane in recorded history was Hurricane Katrina – also from the 2005 season. Hurricane Katrina's storm surge famously broke the levees in New Orleans, causing catastrophic flooding, widespread damage and the loss of more than

Storm damage to a car.

1,800 lives. The financial cost of Katrina has been estimated at over $125 billion, with some estimates putting the damage bill significantly higher. The human toll was particularly heavy on the elderly, who were less capable of escaping the coming storm, and among non-human animals, with 600,000 pets killed or displaced.

Home after hurricane, New Orleans, 2018.

Some of the impacts of catastrophic weather events that are most difficult to quantify can still be seen to be significant. The high media visibility of Hurricane Katrina sparked a media firestorm on the topic of climate change. The impact of Katrina inspired a renewed effort by climate and atmospheric scientists to understand the ways in which climate affects hurricane frequency and intensity, and how a changing climate might influence future storms.[28]

Tornadoes

On the subject of damaging winds, the twisting motion of air is nowhere more visible than in tornadoes. A tornado is a rapidly rotating column of air which is in contact with the ground at its lower end and with a cloud at its top. They can also be called 'twisters' or 'whirlwinds', and sometimes 'cyclones'. While wind is usually not visible to the naked eye, the movements of a tornado are able to be seen, due to the effect of a 'condensation funnel'.

Tornadoes are an international weather phenomenon, occurring on all continents except Antarctica. The United States experiences on average more than 1,200 tornadoes a year; it is the country with the highest number of recorded annual tornadoes in the world, ahead of Canada. These tornadoes frequently occur in a central region of the country known as 'Tornado Alley'. While there is little consensus on the precise boundaries of Tornado Alley, it can be very generally thought to include areas of Texas, Oklahoma, Kansas, Louisiana, South Dakota, Iowa and Nebraska. In the 1920s, tornadoes killed more than

Artist's rendition of a tornado destroying a structure.

three hundred people in the United States annually; however, improvements in warning systems have seen that number decrease to closer to fifty deaths each year.

Tornadoes are associated with extremely high wind speeds, but their winds are difficult to measure accurately. Because of their unpredictable movements and efficient destruction of weather instruments, the wind speeds of tornadoes (also known as 'twisters') are rarely quantified.

Since the 1970s, the impact of tornadoes has been measured by the Fujita scale, which provides an alphabetical and numerical ranking for the event based on the damage caused. The system was developed in 1971 by Ted Fujita, a severe storms research scientist at the University of Chicago. Born in Kitakyushu, Japan, in 1920, Fujita was teaching physics in Tobata, southwestern Japan, in August 1945 when the atomic bomb was dropped on Nagasaki. The city of Tobata itself was the original intended target for this second nuclear strike, but cloud cover over the city on the day of the attack saw the pilot switch to the back-up target of Nagasaki.

In September of the same year, Fujita travelled with a group of students to Hiroshima and Nagasaki to inspect the devastation caused by the nuclear strikes. The trip provided insights that would spark his life's work on destructive events and disasters. While studying the radiation patterns at the site of the bombing, Fujita was struck by the 'starburst patterns' of trees uprooted by the blast. He later used the patterns of devastation left behind by tornadic winds to measure and categorize them into the Fujita scale. The system uses damage to rank the destructive force of tornadoes, due to the problem of being unable to measure tornado intensity while the events are in progress or by appearance. The Fujita Scale begins at F0 and continues up to F12. A classification of F0 reflects 'light damage', such as the uprooting of small trees and the windows of cars and houses being blown out. Tornadoes have been recorded with speeds up to the F5 level of the Fujita scale, which represent damage including the demolition of brick homes and cars becoming mangled after being airborne. Winds reaching an F6 would be associated with an 'inconceivable' tornado according to Fujita, and the remaining

Twister in a field, Wyoming.

measures on the scale, up to F12, were developed to record winds generated by nuclear weapons.[29]

Fujita's lifetime of research into destructive weather events earned him the nickname 'Mr Tornado', and in the 1970s he was instrumental in the development of improved sensing technologies to quantify tornadoes. He was quoted as saying he was interested in 'anything that moves', and this interest saw him build a machine capable of creating miniature tornadoes in his laboratory.[30] Fujita's curiosity about wind saw him greatly advance the understanding of a weather phenomenon called 'micro bursts', which are types of severe downdrafts that create dangerous turbulence for aircraft.

On 3 May 1999, a group of radar trucks (known as Doppler on Wheels) recorded the fastest wind speeds ever measured on Earth's surface, close to Oklahoma City in the United States. Winds of approximately 484 km/h (300 mph) were associated with the Bridge Creek–Moore tornado, making it a category F5 tornado on the Fujita scale. The Fujita scale was replaced by the Enhanced Fujita Scale (EFS) in the United States in 2007. Improvements in the collection and interpretation of weather data had allowed for a greater understanding of the connection between wind intensity and tornado damage. This correlation was reflected in the revised EFS, created by a forum of wind engineers and meteorologists.

Warfare and weather are unruly forces that have shaped the path of human civilization. The uncontrollable nature of warfare and wind have seen both areas linked in human thought to supernatural and divine influences. While the weather can be a powerful ally on the battlefield, its frequent volatility renders futile attempts to anticipate the outcomes of martial conflict. The fortunes of war, it would seem, can shift as easily as the wind.

The unpredictable yet constant threat of catastrophic weather events provides a stark reminder of the often-hidden extent to which the well-being of human civilizations is dependent on favourable climate conditions. Yet throughout the history of measuring stormy weather, scientists, engineers, climatologists and others have been inspired by this changeability to work towards making humans better prepared for these events, and to have a deeper awareness of the complex natural environment in which we live. The need for this research can be viewed as a rare constant feature in a changing world.

5 Trade and Technology

Wind's germinating role stretches far beyond the natural world, in the facilitation of cultural exchange and the growth of global commerce. The famous 'trade winds', blowing towards the equator, are known for their role in aiding economic expansion in the Americas. Wind has assisted nautical explorers and merchants since prehistoric times.

The term 'wind technology' often brings to mind the relatively recent phenomenon of wind farms, but the technological harnessing of wind has ancient roots. Wind technology has been utilized in Tibetan religion to power prayer wheels for thousands of years. Evidence for wind-related technological adaptations can be found in early writings from Persia. In the present day, wind farming is a rapidly expanding international industry – but its growth has not been without controversy.

Despite a less-than-stellar history, the science of weather modification continues to be used to explore solutions to the problem of damaging winds. Geoengineers have been developing technologies that might dissipate the powerful winds found in hurricanes, through the targeted release of air bubbles that reduce water temperatures driving the hurricanes' formation. While wind technology is often used to provide sustainable energy or attempt to mitigate the destructive power of nature, at times, wind technology has been harnessed in the service of warfare.

Trade and wind

A dandelion releasing seeds in a passing breeze.

Since ancient times, humans have been using wind to expand their horizons. Evidence for the use of seaworthy boats stretches back some 50,000 years into the past. The endeavour to capture the wind to power these boats using sails is seen in the archaeological records of Mesopotamian and Tripolye cultures from around the seventh and sixth centuries BCE. Archaeological records of the Tripolye culture (a Neolithic–Eneolithic archaeological culture of Eastern Europe) have been found in modern-day Moldova and Romania. Historical records from the Ubaid period in ancient Mesopotamia (6500–4000 BCE) show evidence for sailing-boat models.[1] These boat models, both riverine and maritime, show some diversity, indicating a variety of types of watercraft in use.[2] Most of the boat models consist of reed-bundles and bitumen coating, and they have been discovered at coastal sites as well as inland.[3] The small, masted Mesopotamian boats are thought to have carried traders between the emerging villages of the Fertile

Crescent.[4] In more recent times, the effects of the 'trade winds' close to the equator have been recognized for their influence on world culture, history and geography.

Jan Porcellis, *Vessels in a Moderate Breeze*, c. 1629, oil on wood.

The 'trade winds' are prevailing easterly winds that circle the Earth near the equator. Trade winds are caused by a complex set of atmospheric conditions specific to their geographic locations. Around 30 degrees north and south of the equator lie the subtropical latitudes. These areas are frequently home to high atmospheric pressure belts and have become known as the 'horse latitudes'. These horse latitudes are associated with clear weather, sunny skies and gentle breezes. The name 'horse latitudes' may have arisen from the actions of sailboats in the region being carried along or 'horsed' on oceanic currents, but the etymology of the term is obscure. One legendary explanation is that it came from the custom of Spanish sailors carrying horses to the Americas by boat for trade.

In the region of the horse latitudes, Earth's rotation moving through space causes air to slant towards the equator in a southwesterly direction in the northern hemisphere. In the southern hemisphere, the same global action causes the air to slant in a northwesterly direction. This phenomenon, the Coriolis effect, is also the force behind the spiralling motion of hurricanes. The Coriolis effect causes the prevailing winds from the horse latitudes to move from east to west on both sides of the equator – these are the 'trade winds'.

The trade winds propelled ships from ports in Europe and Africa across the ocean to the Americas, and the same effect drove vessels from the Americas to Asia. The power of the winds blowing west from the equator to power the development of global commerce led to their being named the 'trade winds' around the time of the early fourteenth century CE. The trade winds supported the development of commercial networks fundamental in shaping the global economy. Along with bringing cargo such as sugar, livestock, textiles and rum to distant shores, the trade winds also powered the transatlantic slave trade that exploded between the sixteenth and nineteenth centuries. The

Ptolemaic world map with twelve windheads, woodcut in Gregor Reisch, *Margarita philosophica* (1503).

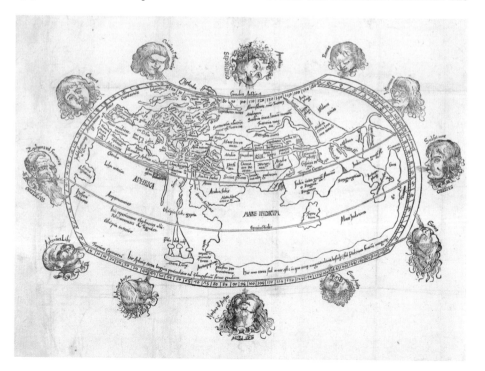

trade in enslaved people radically altered the shape of civilizations across the globe, and its legacy still operates in the present
day.

Wind's guiding force has influenced the shape of human
civilization along with the outlines of the Earth. As described by
the British astrobiologist Lewis Dartnell, the patterns of wind
and ocean currents create 'invisible geographies' that determined
the form of early trade and globalization.[5] The invisible influence
of natural forces has connected previously disparate regions of
the planet, changing the course of human history.

Technology

Long before the power of wind was filling sails in the Age of
Exploration, the motion of air was harnessed with technology. The pairing of wind and technology begins to appear in
the historical record early in the common era, although it is
likely this coupling had its origins in much earlier times. Early
engineering involving wind technology paired wind and water
to create groundbreaking inventions and cultural change. In
ancient China, Han Dynasty engineer and politician Du Shi is
credited with inventing the double-acting piston bellows in the
first century CE. The device used hydraulic power, in the form
of a waterwheel, to operate a piston bellows on a blast furnace.
The invention was used to forge cast iron and is credited with
saving a great deal of physical labour for corvée workers. The
engineer's concern for the labour burdens of working people
earned him great popularity and the nickname 'Mother Du'.[6]
The double-piston bellows was later used in China to pump
petrol for fuelling a type of flame-thrower used in battle around
the third century CE.[7]

As early as 200 BCE, simple windmills were being used in
ancient China to pump water and irrigate crops. Windmills were
also used in Persia and throughout the Middle East, to move
water and grind grain. References to the Persian windmill are
found in the works of 'Ali al-Taburi, the Muslim physicist and
scholar, as well as later writers, who tell the story of the murder of

Umar ibn al-Khattab. Umar is considered one of the most powerful Muslim caliphs in history. He became the second orthodox caliph of the Rashidun Caliphate in 634 CE, shortly after the death of the Prophet Muhammad in 632 CE. According to the story, the Persian slave Piruz Nahavandi (also known as Abu Lulu) assassinated Umar by stabbing him with a hidden dagger. The details of the murder show some variation between tellings, but in several accounts the motivation for the killing is given as Nahavandi turning to violence after Umar rejected his plea for a lower rate of taxation. It is said that Umar had also asked Nahavandi to build him a windmill, as he had heard that the slave had the ability to build powerful machines that harnessed the forces of the wind.

In Sri Lanka, monsoonal winds were used to fuel giant furnaces in which steel was melted for industry. These wind-powered furnaces were the driving force behind South Asia's pre-eminence in steel production in the first millennium BCE. While their use to stoke the industrial flame was at its zenith between the seventh and eleventh centuries CE, recent discoveries have shown the technology dates at least as far back as the third century.[8] British and Sri Lankan archaeologists, working in the dry region of Samanalawewa, found the ruins of 41 wind-powered furnaces in the mid-1990s. These furnaces adapted the prevailing winds to create a dependable draft which was used to keep the charcoal fires sufficiently hot for smelting, thereby creating large quantities of quality steel.[9] The strong monsoonal winds were able to heat the furnaces to over 1,100°C (2,100°F), creating high-carbon steel.

Windmills

Around the fourteenth century CE, windmills became a popular type of power generator throughout Europe. Windmills in this time are most famously connected to Dutch culture, where they had a significant impact. Windmills were used in the Netherlands to drain low-lying regions of land, which was then used for agriculture. With a third of the Netherlands lying below sea level, the power of windmills was crucial for terraforming the country into its current shape. Socially, windmills played a similarly important role in building connections between communities. The sails of the windmills were used for communication between villages. The resting sails were moved into set positions to broadcast the news of a birth, marriage or death. Sections of the sail could be removed to add details such as age or gender.[10]

After centuries of widespread use, windmill numbers peaked by around 1850 CE with the advent of the Industrial Revolution. This resulted in windmills being replaced with steam mills and internal combustion engines. The spread of windmills prior to this period saw medieval Europe become the first large civilization to be run on waterwheels and windmills in place of human labour. This innovation had a transformative effect on society and industry, and perhaps provides an early example of a sustainable industrial revolution.[11]

Aerofoils

The use of wind technology in the form of windmills has a history stretching back hundreds of years. Another type of wind technology, the aerofoil, has formed part of civilization for an even longer stretch of time – and in the case of boomerangs, for tens of thousands of years.

An aerofoil is an instrument which, by design, causes the movement of air to flow differently from one side to another. Through curved shapes, fins, sails or perhaps wings, the aerofoil guides the wind over its surfaces to create an uplift and flight. Types of aerofoil include kites, frisbees and sails. Aerofoils form an important part of aerodynamics in modern airflight.

The boomerang is a type of ancient aerofoil that has become a cultural and national icon. Boomerangs are internationally recognized as symbolic of Australia's First Nations people. The oldest discovered boomerangs in Australia were found in the South Australian Wyrie Swamp in 1973. They are considered to be around 10,000 years old, but it is thought that boomerangs are considerably older than the artefacts from Wyrie Swamp

– boomerangs feature in indigenous cave art, which may be up to 50,000 years old. Aboriginal Australians consider boomerangs to be as old as the continent itself.[12] While it is known that the name 'boomerang' comes from Australia's First Nations people, there is little consensus over the meaning and origins of the word.

The usefulness of aerofoils for hunting was recognized by many early communities – boomerangs were used on virtually all inhabited continents in ancient times.[13] Boomerangs have been found outside Australia in places such as Egypt, India, the Netherlands, Austria and Poland. The nature of the discoveries of these aerofoils sheds further light on their ancient links to human civilization. In southern Poland, an archaeological survey of a cave near the village of Krempachy in 1985 led to the discovery of the tusk of a woolly mammoth that had been shaped to form a boomerang. The device is the oldest known boomerang to have been discovered, dating back to around 23,000 years ago, and its creators are thought to be a community of Upper Palaeolithic people who inhabited the cave and hunted reindeer.

Boomerangs have been used for a variety of purposes, including hunting, fire-starting, creating music and digging. The devices can either be returning or non-returning, and can be found in a variety of shapes, sizes and materials. As well as assisting communities to find food, boomerangs have been used as windborne weapons of war in human conflicts. In 2016 a study attributed the wounds from the skeleton of a First Nations man, who died around 1200 CE, as being caused by a fighting boomerang during intertribal conflict.[14]

Kites

Boomerangs are not the only aerofoils to be used in human conflict. Now generally considered as a benign children's toy, the kite was used as a weapon of war, as well as for numerous other military purposes. Kites have been part of human culture for a very long time, which perhaps gives a clue as to why the exact date and circumstances of their development remain uncertain.

The earliest artistic representation of kite-flying is thought to be in prehistoric cave paintings, dating back more than 10,000 years. While it is generally thought that the history of the kite begins in Asia, the wind-riding crafts seem to have developed in several places among different cultures. Evidence for sacred kites has been found throughout Polynesia and beyond. In the Māori culture of New Zealand, kite flying played a central role in some religious festivals, and kites were thought to connect heaven and earth.

Great Haseley windmill, Oxford.

The earliest known written account of a kite used in battle occurs in ancient China: the famed general Han Xin of the Han Dynasty is recorded using an unmanned aviation vehicle (or 'kite') for the purposes of espionage. During a siege, Han Xin ordered that a kite be flown from his camp to the besieged palace, so that he could calculate the distance accurately for tunnelling. Kites have also been used for psychological warfare, with lights and whistles attached to create an ominous spectacle. Weapons of various types have also been historically fixed to kites to create incendiary bombs. In seventeenth-century Thailand, King Phra Phetracha was said to have tied barrels of gunpowder to kites during a rebellion by the governors of Nakhon Si Thammarat and Nakhon Ratchasima.

While they have largely been replaced by more recent technology, weaponized kites are not yet entirely a thing of the past. Kite flying is a popular cultural pastime in many parts of the world, including in Nepal and Chile. In this sport, participants fly lightweight kites with the objective of cutting opponents' lines, causing their kites to drift free. The practice of chasing the drifting kites to capture and keep them is known as kite running. The practice of kite running in Afghanistan gained global awareness through its representation in Afghan-American author Khaled Hosseini's bestselling book *The Kite Runner* (2003). Kite running was banned in Afghanistan by the emerging Taliban regime in 1994, due to concerns that it distracted young men from religious activities such as prayer. The ban remained in place until the end of Taliban rule in 2001, although some followers of the sport continued the practice in secret despite the ban.[15]

In 2018 the use of incendiary kites in the conflict between Israel and Palestine in the Gaza strip made international headlines. Kites carrying oil-soaked rags or burning charcoal were flown across the Gaza–Israel border by Palestinians protesting the Gaza blockade. The simple weapons set alight fields and forests, drawing swift countermeasures from Israel in the form of airstrikes and fuel restrictions. In the present day, as in the past, the use of kites in warfare continues to be a dangerous business.

Alongside their use in warfare, kites were used in the development of scientific theories and technologies. In 1749 the Scottish team of Alexander Wilson and Thomas Melville conducted the first recorded weather experiments using kites, when they flew airborne thermometers to test atmospheric conditions.

Perhaps the most famous use of a kite in the history of science is the kite experiment of American polymath Benjamin Franklin. On 10 June 1752, Franklin flew a kite in a thunderstorm, allowing him to collect an ambient charge of electricity in a Leyden jar (an antique device for storing an electrical charge). The experiment revealed the connection between lightning and electricity.

Wind and human aviation

From aerofoils to aircraft, the human fascination with flight has a long history. The flight of most modern aeroplanes comes from guiding the movement of air over the craft's wings, creating 'lift', an upward force greater than the downward pull of gravity. This anthropogenic 'lift' means that aircraft can fly regardless of whether the wind is blowing in the plane's location. The 'wind effect' on aircraft still makes a significant difference to the speed and quality of flights, and wind has also played a key role in the development of modern aviation.

On 21 November 1783, a significant step forwards for human flight was made when the first manned hot air balloon success-fully took off for a 25-minute flight over Paris. Developed by brothers Joseph-Michel and Jacques-Étienne Montgolfier, the flight of the manned hot air balloon followed an earlier exhi-bition flight at the royal palace in Versailles, with King Louis XVI of France and Queen Marie Antoinette in attendance. The earlier flight – whimsically named 'Montauciel' ('Climb to the sky') – was crewed by a rooster, a duck and a sheep, and the safe landing of the animal aeronauts saw the king grant permission for human flights.

In 1901 two American brothers, Orville and Wilbur Wright, were in Dayton, Ohio, working on various models of aircraft – and racing to solve the 'flying problem'. Issues with earlier test models convinced the two bicycle manufacturers that what was required was better testing and enhanced pilot control. To increase their testing capacity under controlled conditions, the Wright Brothers built a wind tunnel. The type of simple enclosed wind tunnel that they built had been invented just thirty years earlier, when British marine engineer Francis Herbert Wenham designed and operated the first enclosed wind tunnel in 1871. In the United States, the technology was still uncommon.

The Wright Brothers' wind tunnel was constructed from a wooden box with a square glass window on top for viewing the interior during testing. A fan was belted to a one-horsepower engine borrowed from the Wrights' bicycle shop which provided airflow. The Wright Brothers used the tunnel to measure the forces of lift on drag on dozens of miniature wings, using innovative instruments called 'balances'. The use of the wind tunnel by the Wright Brothers was critical to the successful creation of the Wright Flyer, which made the first controlled, sustained flight of a powered, heavier-than-air aircraft on 17 December 1903. The Wright Flyer was followed by the Wright Flyer II in 1904, and then the Wright Flyer III in 1905. The Wright Flyer III is considered history's first practical fixed-wing aircraft.

As well as guiding the way for modern aircraft, the use of wind tunnels by the Wright Brothers has influenced the design and testing of processes associated with modern flight. Following the use of the Wright wind tunnel, the French engineer and architect Gustave Eiffel built an open-return wind tunnel at the foot of the Eiffel Tower, which he patented in 1911.[16] Eiffel's experience with wind stress in architecture was instrumental in his involvement in the construction of the Statue of Liberty.

Wind tunnels are used by organizations such as NASA to test planes, spacecraft and rockets. The wind tunnels used for testing spacecraft need to be able to mimic alien landscapes, such as Mars. An 8-metre-long (26 ft) wind tunnel in Denmark replicates the dusty atmosphere of the Martian landscape. The air pressure inside the tunnel can be less than one-hundredth of terrestrial sea level, and liquid nitrogen can lower the temperature to −170°c. Fans then blow the remaining atmosphere to test instruments, solar panels and mechanical parts to ensure they can function in the alien conditions.

The world's largest wind tunnel is in California. The NASA Ames National Full-Scale Aerodynamic Complex (NFAC) was constructed in the 1940s and is 24 × 12 m (80 × 40 ft) in size. Its

wind tunnel can provide test velocities of up to 560 km/h (350 mph). The same complex, operated by the United States Air Force, also houses the Unitary Plan Wind Tunnel (UPWT), which is NASA's most heavily used wind tunnel. Over the last forty years, the tunnel has tested every major American commercial transport, countless military crafts, the Space Shuttles and the Apollo, Mercury and Gemini spacecraft. From the beginning of modern human aviation to the present day, from Earth to Mars, wind tunnels have played a critical role in human flight, and in the safety and design of modern aviation.

Wind farming

'Wind energy' involves the use of wind to provide the mechanical power for the movements of turbines. As considered earlier, the power of traditional windmills has been used for hundreds of years to pump water and grind grain, in a traditional type of 'wind power'. In the modern day, the mechanical power of the turbines is used to turn electric generators, making it an alternative resource to fossil fuels.

In 1860 prior to taking office as the sixteenth president of the United States, Abraham Lincoln presciently observed the remarkable potential of wind to provide energy. Indeed, Lincoln predicted that finding the necessary tools to harness this energy might be one of the greatest of all future discoveries.

Lincoln's assassination five years later precluded him from seeing the rapid advance of the wind technology he had envisaged, but developments in wind-harnessing technology had already begun. In the late nineteenth century, the American inventor and philanthropist Charles Brush built the first wind turbine, which converted the power of wind into electricity. Brush's invention played a key part in the rapid development of electrical science during the 1800s, which built on developments from the seventeenth and eighteenth centuries by scholars such as Robert Boyle, Otto von Guericke and Stephen Gray. Brush set up a company called the Brush Electric Co. in 1880 to sell his electric street lighting. This followed the first use of incandescent

Kite flying, Muir Beach, California.

street lighting in 1879, when Joseph Swan lit up Mosley Street in Newcastle upon Tyne for one night. Brush's most enduring invention was what he called the 'wind dynamo'. He built the giant wind turbine in the back garden of his mansion in Cleveland, Ohio, which housed his laboratory in its basement. The automated wind machine was sufficiently powerful that it provided all the power for Brush's home, which was the first in Cleveland to have electricity. Brush commented on the reliability of his wind-powered invention, saying, 'The wind dynamo was built to go for twenty years, and it never failed to keep the batteries charged until I took the sails down in 1908.'[17]

The growth of the wind farming industry in the twenty-first century has placed wind at the forefront of numerous debates, not least those concerning renewable energy, climate change and the management of Earth's natural resources.

Wind farming in the twenty-first century

The rapid expansion of wind farming in the late twentieth century and the early twenty-first century was propelled by the Arab oil embargo of 1973–4. The international 'oil shock' appeared at a time of shifting global fortunes in the energy markets. From the 1860s to the 1970s, the United States was the largest producer and consumer of oil.[18] By 1970 u.s. oil production had hit its peak, but consumption continued to escalate. At the same time, the discovery of new oil fields in the Middle East led to an increase in global production along with global demand.[19]

In 1975 NASA partnered with the United States Department of Energy to develop wind turbines for electric power at an unprecedented scale. The NASA programme developed models of wind turbines that were larger and more economically viable than previous models, paving the way for the wind turbines now used in wind power plants. In 1978 innovations in the Danish wind industry enabled the construction of the world's first multimegawatt wind turbine, further increasing the efficacy of wind farming. This was also the time that wind turbines were set to turn exclusively in a clockwise motion. In 1979 the first modern wind turbine manufacturing company was started in Massachusetts, and in December 1980 the first modern wind farm in the world opened in New Hampshire. In 1991 wind farming moved from land to water with the opening of the world's first offshore wind farm. Vindeby Offshore Wind Farm was erected off the coast of the Danish island of Lolland by the power company Elkraft, after surveying the waters around the island in 1989.

Offshore wind farming was initially met with scepticism by many in the wind industry, due to the difficulty of maintaining farms at sea, and the corrosive nature of the salty environment. However, the offshore industry has developed rapidly in the thirty years since the Vindeby Wind Farm and is considered one of the most likely future candidates for a major stake in the renewable energy industry. Winds are generally stronger when moving across bodies of water than across land, meaning that offshore farms have the potential to draw greater power with more efficiency than onshore farms. Offshore wind farms have faced less opposition from community groups around their placement, although the selection of suitable sites for offshore farms remains a complex business, particularly in terms of environmental impact. Like wind farms on land, the site selection for offshore wind farms involves the consideration of numerous environmental factors. Offshore wind farms need to be situated with consideration for their distance from the shore or coastline, the depth of the water, the nature of the soil on the sea or lake floor and the placement of electrical cables, among other considerations such as the potential for disrupting sea routes used by humans, birds, bats or marine life.

In recent years, the development of floating wind farms has opened the door to new offshore territory for the placement of wind-based power plants. 'Traditional' offshore wind farms are connected to the sea or lake bed by fixed foundations. The fixed-bottom wind turbine is logistically limited to water depths of 50 to 60 m (160 to 200 ft). Without the need for a fixed connecting point, floating wind turbines can be deployed in much deeper water than fixed-bottom turbines, allowing them to have access to more powerful winds – it is estimated that 80 per cent of the world's wind energy is located over water that is of greater depth than can be accessed by fixed-bottom turbines.[20] Greater access to a selection of locations for deployment means that floating wind farms can more easily accommodate the movements of trade and migration sea routes.

While the concept of the floating offshore wind farm has been in development since the 1970s, it was only as recently as 2007 that the first offshore wind turbine was deployed. This prototype, from a technology group based in the Netherlands, was installed off the coast of Apulia in Italy. Ten years later, off the coast of Scotland, Hywind Scotland went into operation – the world's first commercial wind farm using floating wind turbines. The floating turbine technology has proved to be workable, but the increased economic costs associated with the model continue to limit the floating farm's development and deployment. Research is continuing to reduce costs and refine technology to increase the uptake of floating wind farms for future energy needs.

From windswept plains to territory far out at sea, wind power has come a long way in the past hundred years. Yet the future of wind farming may lie even further afield – beyond the final frontier of space. It has been estimated that the energy needs for the entire planet could be met 100 billion times over through the use of space-based satellites that harness solar winds and then beam the energy back to Earth.[21] Solar winds have also been proposed as potential sources of energy for powering future spacecraft. A model for harvesting this valuable solar wind is the theorized Dyson–Harrop satellite. This megastructure would work through 'charging' a copper wire using electrons from space wind. The collected energy would then be transferred back to Earth using an infrared beam. To bring the project from theory to workable practice, further developments in technology and the ability to detect solar winds would be required.[22] With solar wind farming's projected ability to provide extraordinary quantities of sustainable energy, the continued research is a scientific endeavour with much future potential.

The United Nations' Climate Change Conference (COP26) in 2021 saw wind energy feature as a critical component of efforts to keep the increase in global climate warming below 2°C (36°F). A key goal of the international agreement resulting from COP26, the Glasgow Climate Pact, was to promote renewable energy sources as a means to transition away from global dependence on fossil fuels. The proposed acceleration of the implementation of clean power generation, such as wind technology, was adopted as part of the Pact's mitigation strategy to attempt to keep global temperature levels from rising over 1.5°C (35°F), in line with the 2015 Paris Agreement. Although the Glasgow Climate Pact was widely criticized for its ambiguity, some clear targets resulted from COP26 on the further development of wind farms. The Indian government committed to generating 500 GW of renewable energy by 2030, including 140 GW from wind farming. The Global Wind Energy Council (GWEC) assessment of 2022 found that India has risen to rank fourth in the world for installed wind capacity, with 40.1 GW as of January 2022. The country has also become a major hub of turbine component manufacturing and exports. In the future, India's expansive coastline is considered to be highly advantageous for developing a strong stake in the growing global offshore wind market.

In the twenty-first century, the growth and development of wind farming internationally has been exponential. Wind power, like solar power, is considered 'green' energy – it has the capacity to provide renewable energy without producing the polluting emissions that result from the use of fossil fuels. Wind farming has been embraced for its ability to supply 'clean' and sustainable energy at a time of unprecedented demand, but the rapid growth of wind farming has not been without controversy.

A climate of controversy

Wind power is considered one of the leading sources of renewable energy in the modern day. Its ability to offer a sustainable alternative to fossil fuels in the age of climate change and environmental degradation has been a significant factor in the

resource's swift uptake. While it is necessary to find renewable resources for the health of the biosphere, the rise of wind energy to provide more sustainable sources of energy has come with some environmental costs. The powerful turbines used to harness the wind's power have proved hazardous to birds and bats. The loss of animal life through collisions with turbines presents a threat to the preservation of biological diversity. The issue sets up

Rampion Wind Farm, Brighton.

a 'green–green' dilemma, placing competing conservation goals in a conflict not easily resolved.[23]

Wind farms certainly pose a hazard to wildlife; it is generally agreed that wind farms are responsible for the deaths of hundreds of thousands of birds each year – and an even higher number of bat deaths. A study considering data from wind farms in the United States and Europe in 2009 showed that the farms were responsible for the deaths of birds and bats at a rate of between 0.3 and 0.4 fatalities per gigawatt-hour (GWh) of electricity.[24] It is worth noting that the production of other, less sustainable, types of energy also poses a serious threat to wildlife. The same study that found wind farms responsible for 0.3 to 0.4 fatalities per GWh of electricity also found that fossil-fuelled power stations are responsible for about 5.2 fatalities per GWh – almost fifteen times the fatality rate of wind farms.[25]

From an environmental standpoint, the types of animals affected and their conservation status give particular cause for concern. Despite global regulations to reduce the impact of wind farms on wildlife, some farms are positioned in flight paths frequented by threatened or endangered species. In Hawaii, for example, an energy plant off Oahu's North Shore has been found hazardous to the endangered 'Ope'ape'a, or Hawaiian hoary bat. The hoary bat holds the distinction of being Hawaii's only living native land mammal, and the creature is particularly vulnerable to 'predation' by wind turbines. To address the problem, the company that owns the wind farm, Kawailoa Wind, is taking various steps, including changes to wind farm operations, such as the implementation of ultrasonic acoustic bat deterrents on turbines, and measures to increase 'Ope'ape'a numbers by funding programmes to conserve their habitats.

In the Australian state of Tasmania, wind farms are posing a risk to the survival of the Tasmanian wedge-tailed eagle (colloquially known, quite charmingly, as the 'wedgie'). It is thought that around fifty to sixty of the endangered eagles have been killed by wind turbines to date, jeopardizing recent efforts by ecologists to boost the eagles' numbers. In response, Tasmanian regulators are looking to implement locally the remarkable finding of a study into wind farms in Norway.[26] This 2020 study found that by painting every third rotor blade black, the visibility of the wind turbines was increased by reducing 'motion smear'.[27] Bird fatalities were reduced by over 70 per cent by adopting this change, and the benefit was most noticeable among raptors – no white-tailed eagle carcasses were discovered after the painting technique had been applied.

The efforts of local governments, conservationists and power companies in Australia, Norway and the United States have shown that threats posed to birds and bats from wind farms can be mitigated through innovation, research and investing in scientifically proven solutions. Wind energy is a rapidly expanding resource industry and as wind farming technology becomes more efficient, the capacity of wind to meet global energy demands will further increase. The expansion of wind farming means a synchronous move to address the hazards of the technology is required.

While many power companies are proactively working to reduce the dangers of turbine strikes to wildlife, others have failed to meet the minimum requirements of their legally required protective measures. In the Baltic Sea, bats and birds on the major migration route known as the 'Via Pontica' have suffered a high death rate and declining populations due to poor wind-farming practices. Operators of farms in the region have failed to observe the required 'turn-off times' for their machines which would allow the animals to safely pass through the migration corridor. The 'turn-off time' measure has been shown to cost almost nothing and would dramatically reduce the hazards posed to wildlife.[28] The example from the Baltic Sea shows that for the wind energy industry to improve its performance in the sustainability of biodiversity, greater adherence to regulations and development of conservation initiatives are required.

The risks to wildlife posed by the wind farming industry are frequently cited by opponents as a potential reason to embrace other forms of energy production, such as fossil fuel burning. Wind farming does pose a hazard to birds and bats, but the awareness of this threat to biodiversity has resulted in continuing improvements by the industry to reduce animal casualties – even if these improvements and regulations are, at times, imperfectly applied. Less commonly, other potential hazards posed by wind farms are cited as a cause for concern – these include side effects from wind farming on human health and air pollution.

Further wind power controversies

Wind has a powerful and complex influence on human health and behaviour – a topic considered in detail in Chapter Seven. In recent years, however, it has been the influence on health of wind farming, rather than wind itself, that has been the focus of media attention. Wind farms have been noted for producing noise pollution, and concerns have been raised that the health of those living close to the farms may be negatively impacted. Some of the common symptoms anecdotally related to wind turbine exposure include nausea, hypertension (high blood pressure), headaches, insomnia, sensory problems and mental health issues. These symptoms are sometimes grouped together and described as 'wind turbine syndrome'. The wind-farm-related health concerns are often attributed to the emissions of low frequency sounds from wind turbines, which are inaudible to the human ear.

Scientific research has yet to find a strong link between wind farms and the symptoms described by those with wind farm exposure, with a recent research report into infrasound emissions from Finland concluding that these noises were not damaging to human health.[29] The report's authors offered the suggestion that some of the symptoms attributed to wind farm exposure may be triggered by 'symptom expectancy'. While there is currently no empirical evidence for wind farms causing human health problems, the potential for health impacts on communities in close contact with wind farms is a significant issue that requires careful and continuous management. Unwanted noise and visual pollution caused by wind farms may be a cause of stress and worry for neighbouring communities, and a proactive and evidence-based approach to minimizing the impact of farming from the wind industry offers great benefits to both parties.

Early wind farms were noisier and less efficient than wind farms today, and efforts by governments and the wind industry to ameliorate any negative effects that might be caused by the technology have seen the continual development of initiatives around the world to further reduce the impact of the farms. These include the introduction of regulations governing the minimum distance that wind farms can be placed from residential areas, the continual refinement of turbine design to reduce noise, and efforts to build turbines from more sustainable sources.

The need for an improved capacity to recycle old wind turbine blades has only been recognized in recent years – largely because the wind energy industry is still young in comparison with other, more established, energy sectors. While 90 per cent of most wind turbines can be recycled or reused, the turbine blades are less easily managed. Wind turbine blades are typically made of extremely tough but pliable mixes of resin and fibreglass, built to be sufficiently strong to withstand hurricane-force winds. Each blade is around 35 metres (120 ft) long, and the blades' toughness makes them difficult to break down and repurpose. Recently, a recycling start-up company based in Texas has developed a method using the composite fibreglass waste as a manufacturing material, breaking the blades down and pressing them into pellets for use in flooring and walls. As the wind energy industry continues to expand and develop, the need for sustainable solutions for managing its turbines will also increase.

Weather modification

The desire to control or influence the weather has been part of the human experience from ancient times. Many ancient cultures used rituals or magic (or a combination of the two) to attempt to control the weather. In ancient Egypt, magic spells were used to charm the wind, and ancient Mesopotamian spells are directed towards dispersing internal gases.

The magical practice of wind charming is reflected in the Egyptian tale of Setne Khaemwase and Siosiris. The story is best preserved in demotic manuscripts from the Ptolemaic period, but Aramaic versions suggest that its origin is much earlier. The story features the historical figure Setne Khaemwase, the fourth son of Ramses II (1279–1213 BCE), and Setne's son, Siosiris.

The tale involves the two Egyptian royals descending to the Netherworld and discovering the fates of people who have acted justly or unjustly in their earthly lives. Upon ascending back from the Netherworld, the pair are involved in a magical duel with a Nubian magician, acting for the king of Kush. The Pharoah's magician, Horus-son-of-Paneshe, enters into a wizard battle with the Nubian sorcerer. Wizarding battles were very fashionable in ancient Near Eastern literature, also appearing in the biblical Book of Exodus and ancient Sumerian narratives. Among other weather-controlling activities, the Pharaoh's magician silences the wind: 'The Nubian did another feat of sorcery: he cast a big cloud on the court, so that no man could see his brother or his companion. Horus-son-of-Paneshe recited a spell to the sky and made it vanish and be stilled from the evil wind in which it had been.'[30]

In ancient China, music was considered a means to guide the wind towards blowing more harmoniously and guarantee a smooth progression of the annual cycle. Using human-made wind to alter the weather was considered part of good government and was described in the works of Confucius as music's power to 'move the wind and change customs'.[31]

Magic for charming wind from numerous cultures is described by James Frazer in his hugely influential work *The Golden Bough* (1894), including his observations on wind-charming in parts of India, Norway and England. Among many examples, Frazer notes a custom in Austria of offering a handful of feathers to a stormy wind to encourage it to cease.

Apart from the worlds of ancient magic and ritual, human activity has unarguably had a major impact on Earth's weather systems. Fuelled at least in part by anthropogenic causes such as the overexploitation of fossil fuels, climate change has become a growing global crisis in the twenty-first century. The change in climate has exerted an influence on the nature and speed of global winds which remains poorly understood. While the human causes of climate change have been largely unintentional, the ability to purposefully alter weather and winds has long been the subject of human interest. Years of research and experimentation in this area have offered interesting yet troubling results.

In 1947 a collaboration called Project Cirrus between General Electric Company, the u.s. Army Signal Corps, the Office of

Naval Research and the U.S. Air Force made the first attempt to modify a hurricane. The Cape Sable hurricane (also known as Hurricane King) formed north of Panama on 9 October 1947. Over the next few days, it headed northeast, passing through Miami, and appeared to be weakening, making it an attractive option as a test subject for the new cloud-seeding technology that had been under development. Under the direction of the head of General Electric Company, Irving Langmuir, two B-17 aircraft and a B-29 flew into the storm and dropped over 70 kg (154 lb) of dry ice. The goal was not to destroy the hurricane entirely, but to test the technology on a portion of it. Yet following the seeding with ice, the expected response of a weakening storm centre did not eventuate. Instead, the researchers from Project Cirrus watched as Hurricane King suddenly intensified: 'The storm grew stronger, fiercer. To everyone's horror, it then pivoted – taking an impossible 135-degree turn – and began racing into Savannah, Georgia, causing $3 million in damage [$32 million today] and killing one person.'[32]

The newspapers of the time decried the 'weather tinkerers' and their efforts at attacking the storm with 'hurricane-busting' techniques, and threats of litigation were made.[33] Langmuir released a statement saying that he was 99 per cent confident that Hurricane King's hairpin turn was the result of scientific intervention, although other meteorologists were less certain.

Despite the public outcry, Project Cirrus continued to work on weather modification until 1952. The unexpected and destructive pivot of Hurricane King, following its seeding, could be viewed as a kind of harbinger of things to come – in the decades that followed, the history of weather modification would itself take a dark turn.

Between 1962 and 1983, researchers on Project Stormfury carried out experiments to see if seeding with silver iodide could lessen the intensity of hurricanes, and thus reduce their destructive impact. The destructive power of hurricanes created both a strong incentive for and an obstacle to attempts to control them through modern science. The wind energy of an average hurricane is equivalent to half of the world's electricity production for a year. The heat energy released by a fully formed hurricane is comparable to a 10-megatonne nuclear bomb exploding every twenty minutes.[34] A hurricane's destructive potential is primarily dependent on the speed of its winds. During Project Stormfury, U.S. government agencies, including the national weather bureau, attempted to use cloud seeding to weaken the hurricanes' winds. By cooling the water held in the centre of the hurricane, it was thought that the internal structure would be disrupted, and the hurricane would need to redirect energy towards re-forming its eye (the storm's centre), thereby causing a weakening in the force of the winds. To do this, planes were flown directly into the hurricanes, where they dispersed silver iodide. While some early tests did succeed in reducing wind speed, the results proved inconsistent.

Weather as a weapon

Many of the modern efforts to control climate are intended to support the environment or to aid communities struggling to subsist in difficult circumstances. Scientists hope that by reducing the severity of weather systems, some of their destructive effects might be lessened. Yet weather modification has also been used for less benign purposes – where, rather than lessening harmful effects of weather patterns, governments have sought to intensify them. The U.S. government's Operation Popeye used cloud seeding as a weapon during the Vietnam War, to prolong the season of the monsoon winds and increase their intensity. Monsoons are traditionally described as seasonal reversals in prevailing winds, often associated with waves of precipitation.

The use of cloud-seeding technology during the Vietnam War is considered the first recorded weaponization of weather modification in history. Operation Popeye was aimed at securing a tactical advantage for U.S. forces, in the context of their escalating involvement in the conflict. From 1966 to 1972, planes from the United States Air Force would fly over Vietnam, spraying lead and silver iodide into the clouds with the goal of intensifying monsoonal rain. The cloud-seeding programme had a cost of

$3 million each year. The weather-modifying plan was aimed at disrupting the supply lines of the North Vietnamese troops by flooding the Ho Chi Minh Trail and causing landslides, floods and upheaval. Weather modification was a politically sensitive topic, so the operation was conducted in secrecy: planes flew with the stated aim of providing u.s. troops with reconnaissance.

The details of Operation Popeye were only made public through official channels in 1974, and left many unanswered questions – including how the plan may have influenced the eventual outcome of the war.[35] In 1972 the *New York Times* released a story exposing the practice, which was first reported in the *Washington Post* in 1971.[36] The *Washington Post* story noted the problem of 'blowback' and unexpected casualties in using weather as a weapon of war, stating that the American attacks on the monsoons of Vietnam had already shown damaging effects in Laos. The practice of weather control by governments for military purposes was so unpopular with the public that it became known as the 'Watergate of weather warfare'.[37]

Weather modification as an act of war was received poorly by the u.s. public. The general community expressed unease with what was considered to be the government's efforts to 'play God'. The negativity of public opinion led to a rush to pass legislation banning the practice of weather modification as a weapon, and in 1978 a un treaty was signed into law, making the use of weather modification in warfare illegal. This legislation is considered to be sufficiently full of loopholes to be largely ineffective, and weather continues to be viewed as a potential weapon by governments around the world.

The human capacity to harness the wind using technology has broadened horizons in global trade and fuelled the drive towards the sustainable production of energy. Through tapping into the potential for the movement of air to power human activities, the boundaries of territories and societies have experienced continual, wind-powered, change.

Scientific endeavours to provide clean energy and daring attempts to find a means to take flight show the ways in which the natural environment has been both muse and means for extending the bounds of what is considered possible. While these endeavours have provided great benefits to humanity, the technological use of wind through history has not always been positive, as seen in the efforts to control the wind for the purposes of conflict. By exploring wind, trade and technology, we see how the changeable qualities of wind are reflected in the creative and destructive purposes for which the natural environment has been used through human history.

6 Art, Literature and Popular Culture

The invisibility of wind presents a challenge to artists working in visual media. Yet wind has featured in many well-known artworks, with its ephemeral presence at times signalled through divine symbolism. Aeolus, the divine Greek keeper of the winds, is depicted blowing on plants, spreading fertilization. The biblically inspired image of four angels holding the four winds is a popular motif in Renaissance arts, such as in the work of Albrecht Dürer. As well as appearing in visual art, wind plays an important role in music. Wind carries song and wind instruments have been used since prehistoric times, as long as 43,000 years ago.[1] The use of wind to sound bells and chimes holds a protective religious function in numerous ancient cultures.

A further area of popular culture to be explored in this chapter is the presence of wind in sports, such as wind-sailing, and its surprising impact in the world of modern athletics. The 2018 Winter Olympics (an event somewhat ironically powered by wind farms) in South Korea were 'dominated' by wind, which disrupted several major events including alpine skiing and outdoor ice-skating.[2] 'Wind assistance' continues to be debated in the measurement and observation of athletic records. In recreation and artistic expression, wind has companioned human activities from prehistoric times, playing a powerful role in shaping human culture.

Wind in art

The earliest medieval nautical charts contained wind compass roses – geometric drawings showing colour-coded lines that represented various winds and directions. Wind compass roses were also featured on early magnetic compasses, a sign of the primacy of winds in the history of seafaring and navigation. Trade winds were often depicted on early maps used for ocean navigation by arrows showing their prevailing directions, such as in English cartographer Herman Moll's map of the world from 1732. The wind symbols that appear on modern-day weather maps developed from these earlier traditions, with arrows continuing to be used to reflect wind direction, and the addition of feathers on the arrow symbols depicting projected wind speeds.[3]

At other times, wind's invisibility means that it is often the effects of the phenomenon that feature in works of visual art, rather than wind itself. These effects include images of kite

Katsushika Hokusai, *Fine Wind, Clear Weather*, 1830–32, colour woodblock print.

flying, or of instruments of wind power such as windmills. The interplay between the wind and other natural elements features in the works of the Japanese artist Katsushika Hokusai. In the still-life *Poppies in the Wind* (*c.* 1820–34), flowers are pictured bending under the invisible force of wind, the element visibly presenting through its effects. Similarly, in the landscape *Fine Wind, Clear Weather* (*c.* 1830–32), the atmospheric elements of autumn lend a red hue to Mount Fuji – with the effects of the winds again revealing their presence.

An entirely different approach is found in the paintings of Dutch artist Vincent van Gogh (1853–1890). Instead of focusing on the effects produced by wind, the Post-Impressionist artist discovered a way to depict the movement of air on canvas. Physicists at the University of Mexico have demonstrated that several works by Van Gogh reflect the mathematical structure of turbulent airflow.[4] These works include *Starry Night* (1889), *Road with Cypress and Star* (1890) and *Wheatfield with Crows* (1890). The paintings are from the artist's late period, when

Katsushika Hokusai, *Poppies in the Wind*, 1820–34, colour woodblock print.

Vincent van Gogh, *Wheatfield with Crows*, 1890, oil on canvas.

Vincent van Gogh, *Starry Night*, 1889, oil on canvas.

Sandro Botticelli, *The Birth of Venus, c.* 1485, tempera on canvas.

he was experiencing prolonged episodes of mental agitation. The authors of the study suggested that this mental state may explain Van Gogh's ability to accurately represent the quality of turbulence.[5] The pattern of light and shade in these works mirrors the deep mathematical structures of turbulent flow, such as may be found in swirling air escaping a jet engine. To date, Van Gogh is the only artist recognized for reflecting the movement of air in his work with mathematical precision.

The abstract fusing of wind with divine forces is often reflected in art through symbolism. The Italian artist Sandro Botticelli brought Greek myth to life in *The Birth of Venus* (*c.* 1485). The painting is synonymous with Renaissance art and depicts the Greek goddess of love, Aphrodite (whom the Romans knew as Venus), being carried across the sea to the shore, powered by the breath of the west wind deity, Zephyrus. In his arms, Zephyrus holds a nymph, who may be the object of his desire pictured in Botticelli's *Primavera* (dated to around the same period). In *Primavera*, Zephyrus is shown in the act of kidnapping his future bride, Chloris (also know as Flora), a nymph associated with spring and flowers. Both Aphrodite and

Zephyrus are associated with fertility in Greek myth, showing conceptual alignment with the creative force of wind, scattering seeds and germinating plants.

Sandro Botticelli, *Primavera, c.* 1480, tempera on wood.

Wind in music

Since prehistoric times, wind and music have been fused together. As well as supporting human musical endeavours, wind itself is associated with numerous sounds. The noise associated with wind may be ascribed names echoing human and animal sounds, such as wailing, sighing, howling and whispering. The sounds made by wind are generally caused by the vibrations produced by friction as the wind moves against resistance. In this way, the curves of a tunnel or the tangled branches of a forest can be considered as a kind of natural instrument, brought to life through the movement of air.

The natural qualities of wind in making and dispersing sound have been harnessed by humans and animals to create vocalizations and to communicate. The movement of breath,

Musical instruments of the Greeks and Romans, played by people, cherubs and a Pan figure, engraved plate by J. Pass from John Wilkes, ed., *Encyclopaedia Londinensis*, vol. XVI (1819).

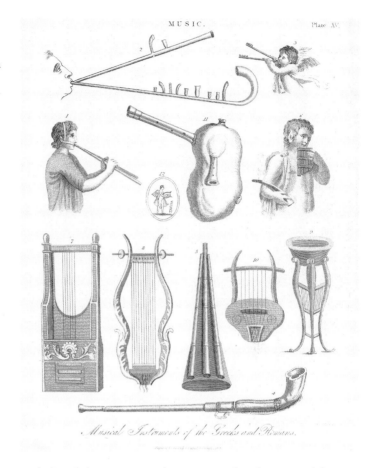

Musical Instruments of the Greeks and Romans.

and the ability to control it, is considered a critical human adaptation in the production of speech. The ability to control the movement of breath has fuelled the development of languages in humans, forming a critical part of the evolutionary process. To speak, and perhaps to sing, required ancestral humans to develop volitional control over their breath. Traditionally, it was thought only humans could do this, and that this ability could partially account for the dominance of humans as a species on Earth. Recent studies, however, have shown that some of our close relatives among the great apes can also exercise breath control. Indeed, it was this ability that allowed Koko, the famous human-fostered gorilla, to play wind instruments.

Seated nobleman with ocarina (vessel flute), Mexican (Jaina), 550–850 CE. The ocarina is used by shamanic practitioners in the modern day to communicate with the spirit world, likely a continuation of an extremely ancient practice.

Wind and breath support the human voice and allow for the production of song, as well as carrying sounds through the air to an audience. The sounds and songs of humans and animals are crucial means of communication between individuals and groups, while also carrying an intrinsic meaning. Debate has raged from at least the time of Charles Darwin as to whether birdsong can be considered analogous to human music, in the sense of carrying an innate meaning as well as creating sound. A recent study of female white-throated sparrows showed the birds had similar neural responses when listening to male birdsong to those of humans hearing music that they enjoy.[6] For many creatures, the wind carries and helps to create the sounds of communication that build connections between group members and allow for melodic expression.

Humans have enhanced their natural capacity for song with the use of specialized tools, with some of the earliest of these being powered by wind. A recent archaeological discovery found the earliest known wind instruments to be flutes made from mammoth ivory and vulture bones, dating from between 42,000 and 43,000 years ago.[7] Early musical instruments are thought to have been used for recreation or religious purposes, and scholars suggest that playing musical instruments may have helped our species of human get an evolutionary advantage over its Neanderthal neighbours.[8] This would be due to the capacity for music to enhance communication within large social networks, which helped prehistoric humans to expand their territory. Along with bones, other items were used in prehistory to create wind instruments. Ancient cave dwellers whose remains were found in the modern-day region of the French Pyrenees have been found to have modified a conch shell by adding a mouthpiece. The shell is thought to have been turned into an early wind instrument around 17,000 years ago.[9]

Wind instruments in use in contemporary Western music include brass and woodwind instruments. Brass instruments, such as the trumpet, French horn, tuba, saxophone and trombone, and woodwinds such as the clarinet and oboe, differ in their usage and materials but involve the movement of breath through the instrument causing vibrations.

Ming Dynasty (1368–1644) ceramic sculpture of an official holding a wind instrument.

While today the instruments are found in orchestras, wood-winds have been played for thousands of years by shepherds and are often featured in compositions with a rustic mood.

The movement of the wind has also been used to create music in the form of wind chimes, a percussion instrument which has been used to give voice to the wind for over a thousand years. There is great diversity in the forms taken by wind chimes, but they generally involve a wind-catching device that is suspended in the air. The bells or chimes can be made of wood or metal. The shape and material of the instruments determine the types of sounds that are made, and the choice of location influences the frequency and types of sounds that will be created. The unpredictability of the wind means that wind chimes can be considered as an example of aleatoric music, or music that contains an element of chance. This unpredictable quality well aligns with the frequent conjoining of wind, change and fate in music and culture – such as in the Cold War anthem 'The Winds of Change', by German rock band the Scorpions.[10]

u.s. Marines with the 3rd Marine Aircraft Wing Band give a performance as part of Marine Week in Seattle, 2014.

Bells, some with clappers to catch the wind, have been found in gravesites across many ancient cultures, such as in Graeco-Roman Egypt, where they were thought to hold magical, purifying qualities. The Roman author Pliny the Elder described the tomb of the Etruscan leader Porsena in modern-day Tuscany as containing pyramids and strings of bells which chimed when the wind blew. Throughout ancient Rome, bells called tintinnabula were hung in gardens. When touched by the breeze, their chimes were thought to have an apotropaic, or protective, purpose, warding away dangerous supernatural forces. The name tintinnabulum comes from the Latin word *tintinnare*, meaning 'to ring'. For greater effectiveness at warding off harm, some of the chimes were fashioned in the shape of phalluses or creatures with erect phalluses, which were thought to avert evil with their vital power. The phallic creatures featured on Roman wind chimes were often depicted with wings. Wings, as noted earlier, are conceptually linked with wind. The ability of wings to harness wind for flight conceptually aligns with the use of wind power to

Wind chimes, South Korea.

Tintinnabulum in the shape of a winged quadrupede phallus, 1st century CE, Pompeii.

bring hanging bells to life and fill the air with randomized music. The belief in the sounding of chimes to bring good fortune and ward off evil is thought to have an ideological link with the use of bells to call the community to prayer in the Christian Church.[11]

Bells that sounded in the wind were found in ancient India from the second century CE, and later in ancient China, where wind bells were placed around the edges of buildings.[12] In Japan, wind bells called *furin* are made from glass, metal or clay. The bells are associated with the craftsmanship of the Edo period (1615–1868), when glass blowing was developing as an art form

Roman bronze phallic ornament, part of a tintinnabulum for warding off evil, 1st century CE.

Rafter finial in the shape of a dragon's head and wind chime, originally attached to Buddhist temple building or royal hall, 10th century, gilt bronze.

Furin, Japan.

in the time of the Tokugawa shogunate. Wind chimes in Japan are thought to have been introduced during an earlier period along with Buddhism in the sixth century CE. The gentle sounds of the *furin* are most prevalent in the hot Japanese summer and are associated with the relief offered by a cooling breeze.

The use of wind instruments from early times shows the remarkable ability of the movement of air to give voice to human-made instruments. In more modern times, wind has been a muse for compositions from many genres, with a variety of compositions featuring wind as a subject. Wind in music, as

Kitao Masanobu
(Santō Kyōden),
*The Evening Wind-
Bell*, c. 1783, colour
woodblock print.

elsewhere, is a common simile for the process of change, and for
the unpredictable nature of life.

In orchestral music, wind is the subject of well-known works
such as Debussy's 'West Wind' from *Preludes*, Book 1 (1909–10)
and Alkan's 'Le Vent' ('The Wind') from Opus 39 *Études*, 'Comme
le Vent' (1857) ('Like the Wind'). A gentle west wind is invoked
by violas in the Summer movement of Vivaldi's *The Four Seasons*
(1725), before a northern gale sweeps in threatening storms. In
Mendelssohn's concert overture *The Hebrides* (1833), a swell of
oboes is employed to mimic a gale. Mendelssohn was said to have

written the piece after being affected by the beauty of the natural world while on a trip to the British Isles. The dramatic intensity of wind is reflected in Chopin's Étude Op. 25 No. 11 'Winter Wind' using swirling scales and arpeggios. In *Dante's Symphony* by Liszt (1857), howling winds signal the arrival of the protagonist, Dante Alighieri, into Hell at the end of the first movement.

'Blowin' in the Wind' by Bob Dylan is frequently listed among top protest songs in popular polls. Written in 1962, the song was widely adopted to promote social justice; 'Blowin in the Wind' was performed by folk singers Peter, Paul and Mary on the steps of the Lincoln Memorial, hours before Martin Luther King Jr gave his historic 'I Have a Dream' speech on 28 August 1963. Although often linked to anti-war sentiments, 'Blowin' in the Wind' held particular relevance for the Civil

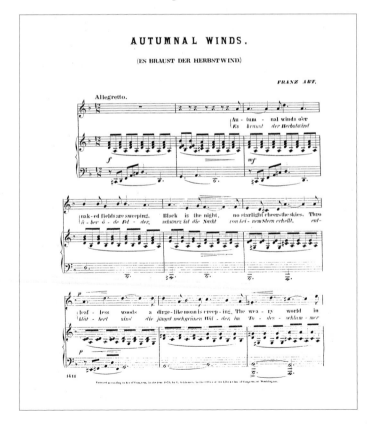

Sheet music for Franz Abt, 'Autumnal Wind' (1872).

Rights Movement – Dylan himself performed the song at a voter registration event in 1963. The song's lyrics, asking how long people can persevere before they have freedom, reflect the struggle of African Americans to achieve civil rights and social equality many decades on from the abolition of slavery in the 13th Amendment of 1865.

Paradoxically, Dylan himself introduced the song when singing it publicly for the first time by saying it should not be considered to belong to the protest genre. Before singing it at Gerde's Folk City in New York in April 1962, Dylan told the audience, 'This here ain't no protest song or anything like that, 'cause I don't write no protest songs.'[13] The statement reflects the broader trend of Dylan distancing his work from attachment to particular movements or issues in the mid-1960s. With his music increasingly being linked to political protests, Dylan was concerned that the polemical interpretation of his music was overshadowing its spiritual and moral dimensions.[14] The artist's frustration with the presentation of his music in some sections of media formed the basis for another song in 1975: 'Idiot Wind'. The chorus of the song describes an 'idiot wind' blowing through every aspect of life and death. Like 'Blowin' in the Wind', the meaning of Dylan's 'Idiot Wind' remains a subject of debate.

Wind in literature and poetry

An anonymous poem, first documented around 1530 CE, describes the unpredictability of the movement of air and the seasons. The poet expresses a yearning for the return of the west wind, which is then connected with the desire to be intimate with a lover: 'Westron wynde, when wilt thou blow, the small raine down can raine. Cryst, if my love were in my armes and I in my bedde again.'

These brief lines are thought to have begun as folk lyrics in the fourteenth century, before their popularity saw them set to music. By the sixteenth century, they had been used as the basis for three masses by John Taverner, Christopher Tye and John Sheppard.[15] The poem juxtaposes the uncertainty of weather with the changeability of life that is part of the human condition. The poet longs for the wind but is powerless to make it blow and similarly cannot control the outcome of matters of the heart. The unpredictable and intangible qualities of wind, evoked in this English poem, are a repeated feature in many written works that creatively engage with the movement of air.

Wind has formed the basis of countless works of poetry. Its mixture of power and intangibility, alongside its frequent connection to unseen supernatural forces, is the theme of a children's poem by the Scottish writer Robert Louis Stevenson. Famous as a novelist and short-story writer, Stevenson also wrote several collections of poetry, including *A Child's Garden of Verses* (1885), featuring his poem 'The Wind'. The poem describes the actions of the wind in lifting kites and birds aloft, and its hidden qualities: 'I saw the different things you did, but always you yourself you hid. I felt you push, I heard you call, I could not see yourself at all.'[16] The poem concludes with a series of questions that show the poet considering the nature of the wind and the source of its unseen power.

The creative and destructive powers of wind are central to Percy Bysshe Shelley's 'Ode to the West Wind' (1820). The Romantic poet defines the wind as both 'destroyer and preserver', as he depicts its violent movement in sweeping away the dead remnants of seasons past, before blowing in the new season filled with 'living hues'. The poet asks for the wind to carry away his lifeless thoughts and bring him new words to spread across the earth, like fertile seeds. This sentiment of renewed inspiration carried on the wind had wide currency in Romantic thought. The poem has been interpreted as referring to the processes of revolution and rebirth, and despite its dark imagery of death and debris, it ends on an optimistic note, with the question, 'O Wind, If Winter comes, can Spring be far behind?' A similar optimism, created by the sense of changing seasons symbolizing an improvement in circumstances, is also found in John Keats's 'The Winter Wind'. In this poem, the reader is assured that for those who have persevered through the darkness of winter, the changing season will bring light, warmth and bounty.

Shelley's use of the west wind to symbolize change can be compared with the similar way the east wind is used in the work of British author Arthur Conan Doyle. In the Hebrew Bible, the east wind is connected with destruction and warfare. The biblical depiction of a dangerous east wind is echoed in the Sherlock Holmes story 'His Last Bow', written by Conan Doyle in 1917. Conan Doyle astutely subverts the biblical image of the easterly's link to warfare to explore the potential for the creation of a new reality through the chaos of the First World War. This idea is raised through a dialogue between the famous detective and his companion, John Watson.

> 'There's an east wind coming, Watson.'
> 'I think not, Holmes. It is very warm.'
> 'Good old Watson! You are the one fixed point in a changing age. There's an east wind coming all the same, such a wind as never blew on England yet. It will be cold and bitter, Watson, and a good many of us may wither before its blast. But it's God's own wind none the less, and a cleaner, better, stronger land will lie in the sunshine when the storm has cleared.'[17]

Having the famously astute character of Sherlock Holmes deliver this oracular speech allows Doyle to give a war-weary public the sense of a possible victory in the near future.[18] The passage further reflects the fusion of wind, change and the divine found in many works of modern and ancient literature.

The post-war winds of change also blow from the east in Australian-British writer P. L. Travers's Mary Poppins series. The first chapter of book one in the six-book series is titled 'East Wind'. It introduces the eponymous magical nanny, who is carried to 17 Cherry Tree Lane by her windborne umbrella.

In the Disney film adaptation of 1964, the arrival of Mary Poppins is preceded by the departure of the conventional line-up of nannies outside the house, blown away on the breeze. Mary Poppins's arrival signals change, and this is symbolically reflected in the use of wind as her means of entering and exiting the lives

Statue of Sherlock
Holmes, London.

of the Banks children. As the nanny says, she will only stay with them 'till the wind changes',[19] which is realized in the book's final chapter ('West Wind'), in which Poppins opens her umbrella and is carried away by the wild west wind.

The pressures of social change following the First World War are reflected in a sense of instability, often related to atmospheric events, in the book. The unconventional aspects of Mary Poppins's character mirror the independence of the children's mother, Mrs Banks, whose ties to the suffragette movement challenge the patriarchal status quo.[20] The portrayal of the two women and their connection to the concept of change and uncertainty that forms a thematic focus of the book reflect the historical social changes of the period with the assistance of wind imagery that signals post-war change.

In Charles Dickens's novel *Bleak House*, the east wind plays a key role in the characterization of the kindly guardian, Mr John Jarndyce. Jarndyce is famously reticent to express his emotions. Instead, he comments on the presence and effects of the east wind to provide a once-removed forecast on his emotional front:

'The wind's in the east.'
'It was in the north, sir, as we came down,' observed Richard.
'My dear Rick,' said Mr. Jarndyce, poking the fire, 'I'll take an oath it's either in the east or going to be. I am always conscious of an uncomfortable sensation now and then when the wind is blowing in the east.'[21]

Nannies blown away by the wind in *Mary Poppins* (1964, dir. Robert Stevenson).

Rick, the young ward, suggests the sensation of unease felt by Jarndyce is rheumatism, in line with the common nineteenth-century medical belief that east winds could aggravate joint pain, as noted in a medical text from 1879: 'The great exciting cause of acute articular rheumatism is cold, especially when the body is perspiring; hence, where east winds prevail, rheumatism is rife.'[22] As in *His Last Bow*, the forecasting of an east wind in *Bleak House* foreshadows danger but also change, which may eventually bring improved fortune through the exposure of secrets and hidden obstacles. The bitter cold of the east wind is further observed in Dickens's *David Copperfield*.

In the works of Conan Doyle, Travers and Dickens, the characterization of the east wind continues its ancient association with altered fates and human conflict. In the Sherlock Holmes stories and Mary Poppins series, wind imagery forecasts social change that is as unavoidable as it is unpredictable for its capacity to shift the established order.

As well as shaping the identity of people, wind is used in literature to characterize settings and build atmosphere. The Santa Ana winds of California are known for their anecdotal influence on human behaviour. The Santa Anas' legendary power to stir erratic behaviour has captured artistic imaginations and seen them feature in numerous cinematic and literary works. The most famous example may be their appearance in a short story by the American-British novelist Raymond Chandler, who depicted the unsettling influence of the Santa Ana winds in the story 'The Red Wind':

> There was a desert wind blowing that night. It was one of those hot dry Santa Anas that come down through the mountain passes and curl your hair and make your nerves jump and skin itch. On nights like that every booze party

H. K. Browne, 'Attorney and client, fortitude and impatience', illustration from Charles Dickens, *Bleak House* (1853).

ends in a fight. Meek little wives feel the edge of the carving knife and study their husbands' necks. Anything can happen.

The description of the Santa Ana winds at the beginning of the story foreshadows the significance of the wind throughout the narrative. The Santa Anas in 'The Red Wind' are a constant and threatening presence. They accompany the hero, Philip Marlowe, as he investigates the strange events surrounding a murder, and their presence has a notably unsettling effect on the story's characters. The intrepid private eye observes the thudding sound of the wind during several climatic moments, such as during the attempt on his life by an assassin. Finally, the interconnected winds and the tension dissolve in the story's final scenes, leaving a 'comfortable' sky.[23] Chandler's description of the Santa Ana winds in the opening sentences of the short story are remarkable for their accuracy and brevity. In his terse style, Chandler encapsulates both the defining features of the wind – hot, dry, blowing downhill – and the effects of the wind on human physiology and behaviour. The nervous sensation accompanying the Santa Anas, described by Chandler, has been thought to be caused by an accumulation of positive ions in the air.

The depiction of the Santa Anas by Chandler typifies their frequent negative cast in artistic works. The Santa Anas are sometimes colloquially named the 'Devil Wind', or the 'murder wind'.[24] In Joan Didion's *Slouching towards Bethlehem* (1968), she describes the unease felt by Californians due to the winds:

To live with the Santa Ana is to accept, consciously or unconsciously, a deeply mechanistic view of human behaviour.

... [T]he violence and the unpredictability of the Santa Ana affect the entire quality of life in Los Angeles, accentuate its impermanence, its unreliability. The wind shows us how close to the edge we are.[25]

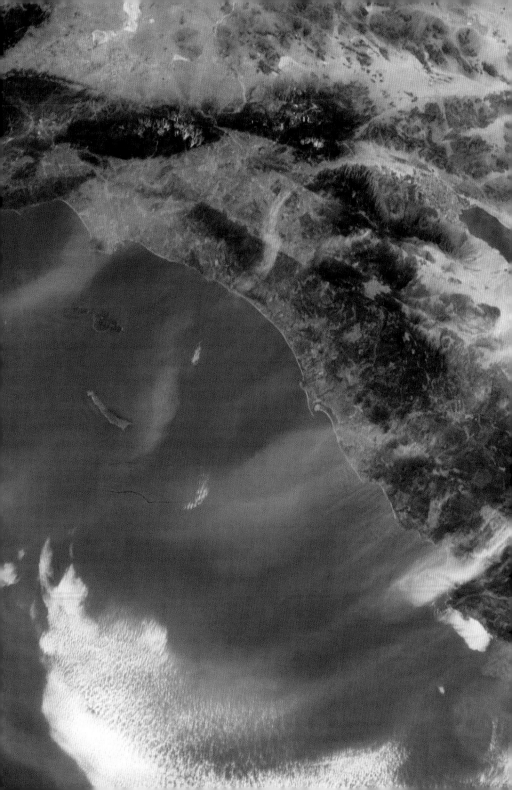

Santa Ana winds,
Southern California,
2002, satellite image.

Wind in literature is used to build a sense of character, story and place. The subtlety of wind, combined with its potency, explains its frequent use in artistic works to bring intangible or divine elements to life on the page, and to provide insights into the causes of chaotic behaviours and events.

Wind on the silver screen

The wind's lack of visibility to the human eye has proved no barrier to its representation in film and on television. While occasionally featuring as an aid to human endeavours, wind in film is often a destructive, even malevolent, force.

Wind is an essentially silent phenomenon, only finding a voice through its effect on other environmental features. Despite this, the sound of howling wind has become a staple of the horror film genre, setting a sinister tone in a similar fashion to the calls of night birds. Many natural sounds, such as gentle rain, are soothing to humans, but the howl of fast-moving wind is connected with a sense of unease. Sounds produced by wind are called aeolian sounds (sometimes spelled eolian). These are produced by the motion of wind as it passes an obstacle. Types of obstacles may include a bare tree branch or the side of a building, but wind can also produce sound when two streams of air converge around an air pocket. Wind that moves through trees with sparse foliage may create 'non-linear sounds', which are sounds beyond the normal range for animals or that change frequently. The non-linear quality of aeolian sound was invoked by the great American poet Emily Dickinson in the late nineteenth century:

> Of all the sounds despatched abroad,
> There's not a charge to me
> Like that old measure in the boughs,
> That phraseless melody.[26]

The 'charge' that is noted by Dickinson created by the atypical sounds of wind is referenced in modern studies that show aeolian sounds increase levels of suspense in audiences when

used in horror films.[27] The sounds associated with the movement
of air, such as howling wind and banging doors or shutters, are
changeable and unpredictable. In film, unearthly howls created
by wind are often used to foreshadow dark or supernatural events
– along with the possibility of sudden change.

Aeolian sounds in the natural world show great diversity.
To recreate the sound of wind for the silver screen therefore
requires selectiveness over the type of wind sound that will best
convey the desired mood to the audience. In the film *Twister*
(1996), for example, the mixture of sounds used to represent
cyclonic winds involved the moaning vocalizations of a camel,
lowered in pitch.

Along with finding the wind's cinematic 'voice', providing visual representation of powerful wind events on screen has
also challenged film producers. In 1938 the film studio Metro-
Goldwyn-Mayer (MGM) bought the rights to L. Frank Baum's
novel *The Wonderful Wizard of Oz* (1900). The first chapter of
Baum's book is titled 'The Cyclone', and it describes a powerful wind event sweeping across the Kansas prairies, lifting the
house holding the protagonist, Dorothy, and her little dog, Toto:

> . . .The house whirled around two or three times and rose
> slowly through the air. Dorothy felt as if she were going up
> in a balloon.
>
> The north and south winds met where the house stood,
> and made it the exact center of the cyclone. In the middle of
> a cyclone the air is generally still, but the great pressure of
> the wind on every side of the house raised it up higher and
> higher, until it was at the very top of the cyclone; and there
> it remained and was carried miles and miles away as easily as
> you could carry a feather.[28]

Recreating Dorothy's trip to Oz inside the cyclone proved
to be the costliest of all the special effects in the fantasy film *The
Wizard of Oz* (1939). Special effects artist Arnold Gillespie first
crafted a 10-metre-tall (35 ft) rubber cone to represent the tornado, but the rubber proved too inflexible to show the twister's

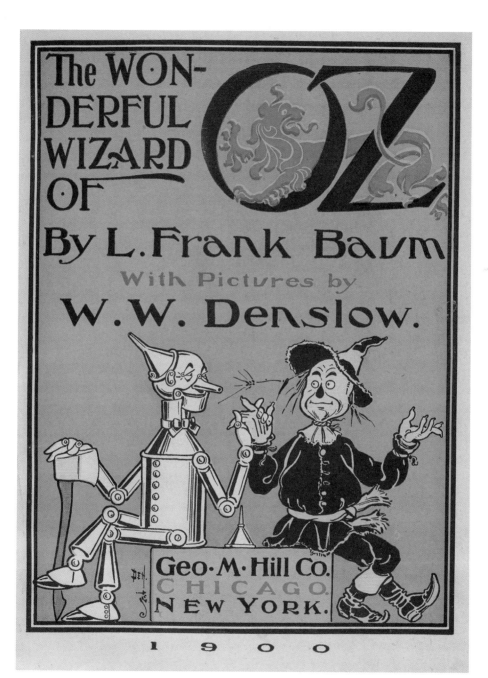

Title page of L. Frank Baum, *The Wonderful Wizard of Oz* (1900), with pictures by W. W. Denslow.

Chapter 1 ('The Cyclone') of L. Frank Baum, *The Wonderful Wizard of Oz* (1900), with pictures by W. W. Denslow.

motion. Inspired by windsocks at the airport, Gillespie then built a long, tapered, muslin sack that was attached to a steel gantry and could be moved by a rod. Air hoses blowing brown dust powder were added to simulate the airborne debris and dirt particles that give visible form to real tornadoes. The tornado sequence and iconic use of Technicolor in the Oz scenes saw *The Wizard of Oz* nominated for an academy award for its special effects – although the film lost out to *The Rains Came* (1939) for its portrayal of an Indian monsoon. *Wizard* was also nominated for the best picture award, which was won by the historical romance *Gone with the Wind* (1939).

Twister

Perhaps the most widely known film with a thematic focus on wind is the blockbuster movie *Twister*. Released in cinemas in May 1996, *Twister* is an American disaster movie directed by Dutch cinematographer Jan de Bont. *Twister* brought the study of climate science and the pursuit of storm chasing to mainstream audiences through an entertaining combination of high-calibre actors, appealing story and award-winning special effects. In addition to grossing almost U.S.$500 million at the box office, the film has had a lasting cultural legacy, including a theme park ride, Twister: Ride it Out, at Universal Studios, and a *Twister* museum in the town of Wakita, Oklahoma, where some of the film was set.

Bill Paxton in *Twister* (1996, dir. Jan de Bont).

The film's screenplay was written by Michael Crichton and Anne-Marie Martin, and the story centres around the efforts of

a plucky group of weather scientists attempting to deploy new tornado-tracking technology in the field. The technology, named 'Dorothy' in a nod to the famous twister featured in *The Wizard of Oz*, will, it is hoped by the researchers, provide better data on the movements of storms and damaging winds, to give affected regions advance warning of dangerous weather events. A sideline to the plot is the romantic reunion of the story's two protagonists, Dr Jo Thornton (Helen Hunt) and Bill 'The Extreme' Harding (Bill Paxton). The team are faced with attempted sabotage from a nefarious group of storm-chasers led by Dr Jonas Miller (Cary Elwes), who is eventually dispatched by a tornado (despite his superior resources).

Twister was notoriously difficult to make. While the movie focused on damaging winds, the film-makers struggled with record-breaking floods and all-pervasive mud.[29] The difficulties of filming in remote locations and inhospitable conditions saw the production beset by accidents and delays. Paxton and Hunt were both temporarily blinded due to burned retinas from overly bright close-up lights. Both Hunt and Paxton needed hepatitis shots after filming in dirty water, and Hunt suffered a further injury to her head during an action sequence which saw her driving a truck through a cornfield.

Twister proved a massive hit at the box office, and its groundbreaking visual effects won awards for bringing the invisible yet deadly force of wind to life. Perhaps the film's most lasting achievement may be found beyond the world of entertainment, in popularizing the study of climate and weather and the pursuit of storm-chasing. Among many meteorologists and weather enthusiasts the film has legendary status and is recognized for its influential 'Twister effect'.[30] The film's success has been cited among several factors, including the impact of Hurricane Katrina, that have 'profoundly altered the meteorological world' and greatly heightened the profile of meteorological and storm studies.[31] The increased visibility of the world of weather science saw an unprecedented growth in undergraduate degree programmes featuring atmospheric science.[32] The visibility of weather studies due to the 'Twister effect' has already been recognized as having a positive real-world impact, as meteorologist Marshall Shepherd observes: '*Twister*'s influence is likely found in research and life-saving initiatives . . . Scientists have suggested that the movie changed public perception of tornadoes and raised awareness.'[33]

Marshall goes on to cite a 2016 article in *The Oklahoman*, which notes that NOAA (the National Oceanic and Atmospheric Administration) selected the University of Oklahoma as the home of the National Weather Center a few years after *Twister*'s release. Both NOAA and the University of Oklahoma inspired the writers of *Twister* with their meteorological research. In the 1970s and '80s, researchers from NOAA designed an innovative device, TOtable Tornado Observatory (TOTO), which was intended to collect weather data from inside a tornado. Teams from NOAA and the University of Oklahoma attempted to deploy the technology in a similar way to that depicted in the film but were unsuccessful in scoring a 'direct hit'.[34] TOTO was retired in 1987, but the legacy of the scientific effort lives on in *Twister* and its following. A reference to the project is seen in the research team member Dustin (a scene-stealing Philip Seymour Hoffman) wearing a red University of Oklahoma baseball cap.

Sharknado

In 2013 a made-for-television film called *Sharknado* became a social media phenomenon, attracting millions of viewers and spawning a film franchise. *Sharknado* explores the science-fiction premise of what might happen if an unusual weather event caused sharks to rain from the sky in urban areas.

Of course, fish, frogs and other small creatures falling from the sky have featured in artistic works since ancient times. In this way, *Sharknado* can be seen as a modern reimagining of an ancient phenomenon found in the literature of ancient Mesopotamia and the Bible – perhaps most memorably in the raining frogs from the Book of Exodus. Ancient authors such as the Roman naturalist Pliny the Elder and Graeco-Egyptian Athenaeus of Naucratis recorded seeing the unusual type of rain involving

falling creatures in their historical works. In recent years, torrents of falling animals have made news headlines around the world, including a fish rain shower in a Sri Lankan village in 2014.[35]

As with *Twister*, *Sharknado*'s cultural impact included influencing public engagement with storm and meteorological studies. The fictional premise of the film was widely referenced on social media to raise awareness on important issues such as

Cloudy skies and shark-filled water spouts in scenes from *Sharknado* (2013, dir. Anthony C. Ferrante).

Waterspout,
Milwaukee County,
Wisconsin, 2011.

storm safety, climate change and weather. The Red Cross used a screening of *Sharknado* to promote its tornado planning and First Aid apps, and the National Weather Service in Kansas tweeted using the hashtag 'Sharknado' to educate its audience on how to build a tornado preparedness kit. Researchers working in the field of weather science used the premise of the film as a platform to encourage community awareness of storm science, by investigating the plausibility of flying (killer) fish.

In the film, the eponymous fish are lifted from the ocean by powerful waterspouts, resulting in flying schools of rampaging sharks. While the film's outlandish premise adds to its light and humorous tone, scholarly processes have been used to assess the scientific accuracy of a *Sharknado*-style event. To test the premise, the first question is 'what kind of tornadic mechanism is at work in lifting the cinematic sharks?'

TROMBE MARINE

Meteorologists generally distinguish between two types of waterspouts: fair-weather waterspouts, which are made up of rotating, cloud-filled wind, and tornadic waterspouts, which are tornadoes that have formed over water, or which have moved from land to water. The presence of stormy conditions in the Sharknado films suggests that the cinematic shark-raining phenomenon can be attributed to tornadic waterspouts. The next question is 'could powerful winds cause giant sharks to rain down from the sky?' Powerful winds can lift land-based items of great weight or those that are strongly attached to the ground. On the Enhanced Fujita Scale, used to measure the intensity of tornadoes, the highest category of known tornado is EF5. At this point on the scale, 'incredible damage' may occur, including the possibility of cars, trucks and trains being lifted by the winds and thrown distances of up to 1.6 km (1 mi.). The famous American Tri-State Tornado of 1925 is thought to fall into the 'incredible damage' category, with stories told of entire houses being lifted from their foundations. The Tri-State Tornado was

Illustration of a waterspout, in Charles Delon, *Cent tableaux de géographie pittoresque* . . . (1881).

considered the longest continuous tornado in recorded history until December 2021, when the unprecedented weather event known as the 'Quad-State Tornado' may have set a new record for the length of the tornado's track, at 400 kilometre (250 mi.)[36] Most tornadoes are rated in the lowest categories of EF0 and EF1 on the Enhanced Fujita Scale, with fewer than 1 per cent of twisters rated at EF5.

A climatologist consulted by the *Los Angeles Daily News* has theorized that a sufficiently massive EF5 waterspout could potentially lift a great white shark from the sea and carry it to land.[37] The movie's setting in California provides a good fit for the unlikely events of the story, as the area is considered the tornado and waterspout capital of the United States. While massive waterspouts of the type required to create storms of sharks raining from the sky are happily scarce, climatologists have noted the potential for climate change and global warming to lead to an increase in extreme weather events – including super waterspouts like those seen in the film.

Wind in sport

Sport in the modern day is an area of international cultural significance, as well as being big business. The focus on achieving previously unmatched records of human performance in a variety of sports has made wind assistance a particularly controversial topic in modern athletics. A favourable wind direction can make the difference between winning and losing, and between being the best on the day, and the best of all time. While every sport is affected by wind to some degree, for a minority of sports, wind itself is the main event. 'Wind sport' is a category of sport driven by wind power, often through the use of an aerofoil (such as a kite or sail). 'Air sports', such as parachuting or hang gliding, also often employ aerofoils, such as parachutes, but are considered less focused on wind than air (although a rather nebulous categorization). While sports such as sailing or hang-gliding are widely known, wind sports include less high-profile pursuits such as snow-kiting, involving riding a snowboard or skis across ice

or snow propelled by a kite, and sail biking, where a kite is used to provide acceleration to a bicycle.

Windsurfing, a wind-powered pastime.

A modern pastime with ancient origins, competitive sailing is an international sport requiring strategy, fitness, seamanship and the ability to read the wind. The sport of sailing is often thought to have emerged in the Netherlands around the seventeenth century, as it was there that the English monarch Charles II became fascinated with the yachts that were kept close to the water as a normal part of life for the wealthy class.[38] The Dutch gifted Charles II a yacht named the *Mary* in 1660, the year of his ascension to the English throne. British shipbuilders were soon at work adding new yachts to the king's collection – with the British versions routinely armed with guns – as the king would keep one or two for leisure activities, while the remainder of the fleet was employed for State purposes.[39]

Sailing for pleasure had been a pastime of royalty even prior to the seventeenth century: the legendary king of Norway

Harald Fairhair was said to have given a beautiful ship to the English king Athelstan (known as 'Athelstan the Good' to the Norwegians). Robert Bruce, the king of Scotland, frequently sailed a small vessel along the river Clyde for leisure, and Queen Elizabeth had a pleasure ship built in 1588 called the *Rat of Wight*.[40]

Sailing for pleasure continued to increase in popularity over the centuries, and in 1720, the world's first known yacht club was founded in Cork, Ireland (although there is some competition for the title from the Neva Yacht Club of St Petersburg, Russia, founded in 1718 by decree of Tsar Peter the Great). The Royal Cork Yacht Club was founded by the great-grandson of one of the courtiers of Charles II. With the establishment of yacht clubs, yachting soon developed into a popular international sport. In 1896 sailing debuted as an Olympic sport, with a regatta planned to take place at the summer games in Athens – although a lack of suitable entrants saw the event cancelled. The early Dutch influence on sailing can still be seen in the etymology of many words associated with the sport: The word 'yacht' comes from the Dutch word *jaghte*, meaning 'light sailing vessel', which developed from *jaghtschip*, meaning 'fast pirate ship'; the word 'cruise' evolved from *kruisen*, meaning 'to cross'.

Along with sea and land sailing, windsurfing is one of the best-known wind sports in the modern day. The sport experienced an explosion of popularity in the second half of the twentieth century, particularly the 1970s and '80s. While considered a 'modern' sport in popular culture, as with its cousin, surfing, the origins of windsurfing can be traced back a long way into the past. Windsurfing is thought to have its origins centuries ago in the islands of Polynesia and has been an Olympic event since 1984.

The 100-metre dash has been a permanent part of the Olympics since 1896, with women joining the field in 1928. A sporting event in which every fraction of a second counts, the dash is especially vulnerable to wind interference. At the same time, the stakes of the race are particularly high; experts call the Olympic 100-metre dash 'the most lucrative ten seconds in sport'.[41] The winner of the men's event can expect to make tens of millions of dollars in sponsorships and endorsements. After winning the gold medal at the Beijing Olympics of 2008, Usain Bolt's sponsorship with Puma alone was thought to be worth £21 million.

In the 100-metre dash, a strong headwind will slow the competitors down, while a fast tail wind can help set new records for speed. The dash is one of a handful of track events with a legal limit for wind assistance – this limit is 2.0 m/sec (6½ ft/sec). Above this limit, the athlete's time is ineligible as a national or world record. The dash, and other events that fall under the legal limitations of wind assistance, are grouped together, as they involve the competitor moving in a single direction. In the 400 metres or 800 metres events, conversely, competitors will experience the prevailing wind blowing to their fronts as well as their back, evening out any potential advantage.

On 16 July 1988, Florence Griffith-Joyner (known colloquially as 'Flo-Jo') ran the 100-metre sprint in a time of 10.49 seconds. This extraordinarily quick time made her the fastest woman sprinter of all time, a record that has stood for over three decades. The record was set at the u.s. Olympic Trials in Indianapolis, and Griffith-Joyner would go on to win the Olympic gold for the 100-metre, 200-metre and 100-metre relay in the Seoul Summer Olympics of the same year (as well as silver in the 200-metre relay). Griffith-Joyner's status as the fastest of all female sprinters is not in doubt – the prodigious American ran similarly speedy times at various meets around the time of the 1988 Olympic Games. But in a sporting world in which every fraction of a second counts, the ephemeral element of wind has created controversy around the record time of 10.49 seconds.

The official wind measurement recorded by the anemometer at the Michael A. Carroll track at the Olympic trials was 0.0 m/sec. However, the weather conditions forecast for that afternoon featured gusty winds, and it is widely believed that the anemometer gave a false reading. A report commissioned by the International Association of Athletics Federations (IAAF) estimated that the genuine measure of wind assistance that day would have been somewhere between 5 and 7 m/sec (between 16

100-metre dash at an event held by the American veterans organization Wounded Warrior.

and 23 ft/sec), well above the legal limit for a world record. This view was based on similar wind readings of around 4 m/sec (13 ft/sec) for the men's triple-jump track which ran parallel to the women's sprint track and was oriented in the same direction.[42] Analysis following the event has raised the suggestion that the wind speed gauge was not plugged in, although the evidence remains uncertain.[43]

Although arguably an ally for Griffith-Joyner's Olympic goals in 1988, wind proved an unruly force at the 2018 Winter Olympics hosted in PyeongChang, South Korea. The 2018 Games were powered entirely by wind, with the region of PyeongChang home to wind turbines that are part of South Korea's plan to boost its supplies of renewable energy. While the powerful winds of the area are useful for powering turbines, in some of the Olympic winter sports the winds caused delays and disruptions. The howling winds unnerved competitors at the ski jump and caused significant reordering of the schedule for the downhill slalom events. The winds also caused minor injuries to spectators from flying debris, as tents and fences were uprooted.[44] High winds in the women's snowboarding events saw most competitors unable to complete their runs, with Australian snowboarder Tess Coady injuring her knee after a sudden wind gust caused her to crash during a practice run.[45] The effects of wind at the 2018 Olympic Games illustrate how even years of precise human planning and training can be upended by the power and unpredictability of natural elements. At the same time, the Olympic ideals of global cooperation and striving for improvement are worthwhile considerations in the efforts of humanity to meet the challenges of living in harmony with the natural environment. This spirit is perhaps reflected in PyeongChang's use of wind energy to provide the first modern Olympics powered by renewable sources.

Whether seeding the artistic imaginings of poets and writers of the Romantic movement, or lent symbolic form as the fertile force of the breeze in Renaissance art, the invisibility of wind has proven no bar to its appearance as a powerful presence in a range of artistic works. At the same time, its hold on the artistic imagination has had surprising real-world effects, in its ability to build awareness and interest in the fields of meteorology and storm science. An ever-present influence on the sporting world, wind has the potential to lift human endeavours to new heights, while at other times it reveals human limitations in the face of exposure to the natural elements.

7 Wind, the Environment and the Future

Wind has been connected to human mood and behaviour for millennia; yet the modern understanding of the relationship between weather, health and behaviour remains incomplete. In Switzerland, sleeplessness and irritability have been linked to the appearance of the warm, dry foehn wind. Across the globe, countless distinctive winds known to local communities are famed for their medicinal or behavioural influences. Of course, it is not only humans who feel the effects of the wind: the effects of wind on animals are a rapidly developing area of study.

As well as considering the influence of wind on behaviour, this final chapter also considers wind in the future – from wind power to the potential effects of human behaviour on wind. The connection between wind and humanity continues to develop dynamically, with recent scientific and technological advances providing a glimpse into the future of wind on Earth – and beyond.

Wind and animal behaviour

Animals rely on wind for a habitable atmosphere on Earth. As well as making Earth's surface sufficiently temperate for life to flourish, wind plays a critical role in distributing seeds and pollens, spreading this life across the planet. Yet wind can be destructive as well as creative. Giant sandstorms and damaging tornadoes provide strong and visible evidence of the potency of wind, and the movement of air can also wield an influence in subtler ways.

Autumn breeze in Taipei.

The appearance of birds soaring on air currents is a quotidian reminder of the interdependence of living creatures and the power of wind. Flying birds take to the skies by harnessing the power of lift, which is an upward aerodynamic force created by differences in air pressure generated by an aerofoil – in this case, the bird's wings. The curved shape of the top of the bird's wings causes air to move more swiftly above the wing than below, and the resulting difference in pressure lifts the wing (and bird) up. Birds can flap their wings to create thrust and increase the movement of air around their wings. The physical effort of flapping can be avoided by soaring birds, such as hawks and pelicans, who can maintain their flight riding on wind currents – sometimes for days at a time.[1]

As well as providing lift, recent research has found that bird wings act as a kind of natural suspension system.[2] In windy conditions, the shape and composition of birds' wings can absorb powerful gusts, allowing them the necessary precision of movement to evade predators or land safely. By virtue of the wing's design, the dampening effect in gusty conditions occurs

Seabirds in windy conditions, Falkland Islands.

automatically, reducing the energy that birds require to adapt to windy weather. While doubtlessly beneficial for many birds, the rough conditions experienced by some seabirds test even the most aerodynamic design.

Wind affects seabird behaviour more than any other environmental element.[3] Seabirds must adapt to living by the ocean, where wind chill can cause temperatures to plummet, and high winds affect nesting and hunting. A 2019 study showed strong winds can prevent cliff-dwelling birds from accessing their nests and caring for their young hatchlings.[4] Landing on the narrow ledges of cliff faces requires delicate control from the birds – a clumsy landing risks dislodging nests along with their fledgling contents. The study showed that, in near-gale-force winds, only 20 per cent of landing attempts by the birds succeeded. This figure contrasted with a 100 per cent landing success rate for the birds in still conditions. The scientists concluded that the presence or absence of winds and their strength had significant relevance for the choice of nesting locations by cliff-dwelling birds.

Wind is not only a concern for the nesting behaviour of seabirds. Our closest known relative, the chimpanzee, has also been shown to take account of the strength of local winds while nesting. A recent study of chimpanzees living in Senegal and Tanzania showed that the animals responded to windy conditions by building nests deeper among tree foliage with stronger structural supports.[5] Great apes are the only primates known to build nightly 'nests', or sleeping platforms, where they bed down to rest. This is a complex behaviour that involves the use of tools, selection of materials and the ability to respond to changing environmental conditions and potential threats. The authors of the study reflected that the ability of chimpanzees to respond to windy conditions by building more secure nests was a skill likely shared with early human ancestors:

> Given we know that all great apes build nests, and that many early hominins retained adaptations for tree-living such as feet that could grasp onto branches or food, it is likely that

they also built varied nests. This would have helped them adapt to a changing landscape and an unpredictable climate during key periods of evolution.[6]

In chimps, the ability to build nests was shown to involve learned behaviours. If this finding can be plausibly extrapolated to early humans, it reflects the close relationship between humans, environments and communities from a very early time in human history, when the presence of wind likely influenced the shelter-building behaviours of distant ancestors of modern humans.

Wind has been a powerful agent of natural selection in the evolutionary development of many species. A recent example of this trend is found in a study on Neotropical lizards (tree-dwelling reptiles with compressed bodies and long, whiplike tails). The study explored the changing forms of lizards in South and Central America and the Caribbean, to consider how they might be affected by exposure to hurricanes. Lizards with bigger toepads were found to be more common in areas that saw frequent

Chimpanzee nest, Kenya.

hurricane activity. Lizards with larger, grippier feet appeared better able to cling to points of security during the powerful winds brought on by hurricanes, while lizards with smaller toe-pads were blown away. The researchers demonstrated that the selective advantage of large toepads was carried into later generations of lizards who had not been exposed to hurricanes, showing the evolutionary impact of extreme weather events. The increase in extreme weather events, intensified in recent years by climate change, may mean that the short, sharp evolutionary impact of intense weather could be a growing trend. Intense weather events may prove to be overlooked drivers of significant biological change.[7]

Future studies will likely assist in building greater awareness of this evolutionary phenomenon, as noted by one of the study's authors, Colin Donihue, in the *New York Times*: 'I think we'll see more and more that there are other species whose evolutionary histories, and evolutionary futures, are impacted by survival of hurricanes.'[8] While the exact contours of future events and evolutionary trajectories remain undefined, it is clear that the invisible shaping force of wind will continue to play a critical role in directing the course of animal life on Earth.

Predation

Recent research has observed the strong yet complex effect of wind on predator–prey interactions, an area of science that is currently poorly understood.[9] The relationship between predators and prey in an ecosystem is influenced by numerous environmental factors, and the modern understanding of the role of wind in these relationships is still developing. Despite the dynamic state of scholarship, wind is known to play a complex and important role in the relationships between predators and their food webs. Animals that rely on alarm calls to signal the presence of a dangerous predator, for example, face increased risk in windy weather due to impeded call transference. American pika birds are less active and make fewer calls in the presence of high winds.[10] Cockroaches have sensory appendages that are extremely sensitive to even slight changes in airflow: their ability to sense wind created by the movements of predators helps them to detect and avoid attacks.[11]

Part of the challenge of observing the influence of wind on predator–prey interactions is that while there have been few scholarly analyses of the topic, the evidence resulting from the available studies often appears contradictory. In some studies, and in certain species, increased wind results in a rise in predation. In other studies of different organisms, the reverse may be true: some predators faced with windy conditions are found to be less successful in capturing prey. The diversity in the observed influence of wind on hunting behaviours is likely due to the subtlety of the subject – wind appears to have an influence on animal behaviour that is unique to different environments, conditions and circumstances. The range of influences that wind has on behaviour demonstrates the intricacy of interactions between species and their natural habitats.

A 2017 study showed the clearest impact of wind on predators is found in the areas of detection, locomotion and disturbance.[12] In the area of detection, wind can conceal or reveal the presence of predators to their potential prey. Predators whose hunting success may hinge on wind direction range wildly in size and geographical distribution – everything from packs of lions hunting buffalo in Kruger National Park, to predatory mites pursuing herbivorous mites along the leaves of lima beans.[13] In the case of lions, hunting upwind has been found to significantly decrease the effectiveness of the hunt, in contrast to hunting in still conditions. The influence of wind on animal behaviour extends even beyond the interactions of predator and prey – hyenas have been found to wait downwind of adult male lions for their chance to scavenge the remains of a kill.[14] This cautious behaviour has been shown to improve the survival rates of the hungry hyenas.

In the area of locomotion, wind has a significant influence on attack and escape behaviours in predators and prey. The effect is commonly seen in aerial predators, such as eagles or falcons. Generally, wind improves the hunting success of aerial predators by giving them increased manoeuvrability.

The final area of predation often influenced by wind is habitat or environmental disturbance. Wind that causes the disturbance of habitats generally has a negative effect on predation. An exception has been found in some species of crabs, which avoid climbing trees during windy conditions and are therefore prevented from accessing shelter. These grounded crabs make easier prey for other crabs.[15]

Beyond the hunting patterns of animal predators, wind itself has been shown to take a toll on animals surrounding wind farms. The ability of wind farms to disrupt the ecological balance of their surrounding environment has been a serious concern since the earliest industrial wind farms took shape in the 1970s. A recent study showed that the presence of wind turbines in the richly biodiverse Western Ghats mountain range of India resulted in birds becoming more scarce and hunting less frequently in areas more densely populated by wind farms. The avoidance of these areas by preying birds saw an increase in the population size of lizards usually preyed upon by birds in these areas. Interestingly, the researchers further noticed that the lizards 'had much lower stress levels' than their relatives who lived further from the wind turbines, providing an unexpected windfall of relaxed reptiles around the farms.[16]

Ocean predators too are affected by the movement of wind. Preliminary findings from researchers studying shark attacks suggest that weather conditions may play a role in increasing the risk of a shark attack on a given day. Specifically, the presence of a sea breeze may have an influence on drawing sharks closer to the shoreline.[17] Much scholarly work has been carried out to discover what role environmental factors play in shark attacks on humans, with the most commonly studied element being water temperature. Along with humans, many other creatures need to concern themselves with evading predators in the ocean. Wind is among several elements that play a significant role in the frequency of shark attacks on fur seals around Seal Island in South Africa. Studies have shown that attacks from white sharks on Cape fur seals significantly increase when a north wind is blowing.[18] While it is unknown why wind direction

Fur seals on Duiker Island (also known as Seal Island), Cape Town.

might influence sharks' hunting behaviour around Seal Island, it is thought that prevailing environmental conditions could make the seals easier to detect.[19] A further study on Seal Island found wind had a significant influence on shark feeding behaviour on whale carcasses during migration. In still conditions or during light winds, few sharks (and sometimes none) were found feeding on whale carcasses. When winds were high and blowing towards the coastline, many sharks were present. The direction of the wind as well as its speed were strongly linked to the sharks' scavenging habits, leading to the conclusion that 'highly sensory consumers' such as sharks are drawn to the whales by chemical and scent signals carried by environmental elements.[20]

Wind and human behaviour

The exponential growth of technological developments in the twenty-first century provides unprecedented opportunities to view the natural world from new perspectives. At the same time, the advance of technology has been frequently decried for its capacity to create a sense of distance between humans and their environment. Yet despite the increasing presence of technology in modern lives, human animals remain intrinsically linked to the natural world. Throughout history, humans and wind have been actively entwined. Wind has changed the course of civilizations, the evolutionary paths of many species and the shape of the planet on which we live.

The movement of air – from 'outside' winds in the natural environment to the 'inside' winds of breath – connects humans to the environment, offering a 'window into the relationship between humanity and the natural world'.[21] The interplay between internal and external winds in health has long been recognized, from the concept of wind in traditional Chinese medicine, to the healing practices of indigenous hunter-gatherers in southern Africa.[22]

The intricacy of wind's effects on human well-being means the topic eludes attempts at neat classification. As we have seen, 'wind' is used as a catch-all word reflecting a variety of movements of air, many of which have been connected to distinct aspects of human health. This creates obstacles for assessing whether the influence of wind might be positive or negative – the effect of wind depends greatly on its character and context and can even vary depending on individual circumstances. We will first look at general trends in wind and human behaviour,

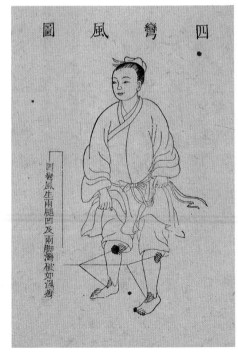

Chinese woodcut illustration of 'Four joints wind' (*siwan feng* – eczema on the ankle joints and behind the knees), from *Yizong jinjian: Waike xinfa* (The Golden Mirror of Medicine: Essential Knowledge and Secrets of External Medicine), early 18th century.

and then move to 'ill winds', before discussing the winds considered to be beneficial for well-being.

Weather and wind exert an often-unseen pull over human behaviours. This pervasive sway can be found in even the most analytical activities. Several studies have shown the significant influence of weather on international stock markets.[23] There is evidence for a change in market behaviour due to humidity and temperature in India, and for rain dampening market activities in Ireland. The effect of wind on market performance is as yet not well understood; the limited number of studies to date means the nature of the effect is tied closely to the scope of each study, rather than revealing broader trends.

In Chicago, known as the 'Windy City', markets have demonstrated a change in the bid–ask spread in response to differences in wind speed.[24] (The 'bid–ask spread' is the amount by which the ask price exceeds the bid price for an asset in the market.) Windy days have resulted in a wider bid–ask spread in Chicago, perhaps representing greater risk aversion in the market. A study from 2016 analysed the Israeli stock exchange and weather and found low wind speeds were connected to low yields, and high wind speeds resulted in higher returns. Windy conditions have also been observed to accompany decreased stock returns in a study of eighteen European financial markets.[25]

A well-known study from 2002 measured the influence of weather on the market in New Zealand. The distinctively windy nature of the city of Wellington, home to the New Zealand Stock Exchange, is said to result in the occasional appearance of seagulls 'seen flying upside down'.[26] Very windy conditions in Wellington bore a significant correlation with lower stock returns than less windy weather. The authors of the study considered whether the distinctive weather in Wellington made for a unique influencing factor on the market, which might explain the different influence of wind on the market in New Zealand compared to other global studies. Taken together, the research shows that wind has an effect on stock markets internationally, but the nature of this effect is diverse, and its mechanisms are not well understood.

The influence of windy weather is seemingly not only felt by investors. Increased windiness, and sudden changes in wind speed, have been statistically correlated with an increased incidence of numerous crimes, including domestic violence and homicide.[27] Interestingly, the incidence of crime relating to wind shows some variation in terms of setting. While crimes committed in interior, private spaces increase with the wind speed, assaults and batteries in exterior, public spaces show a decrease in frequency.[28]

Wind, weather and mood

Anecdotal evidence for wind and weather affecting human mood is ubiquitous. Children are thought to be less well behaved in windy weather, and there is a traditional view that women are more likely to give birth in storms than in fine conditions.[29] Headaches and sore joints have been connected to windy conditions in ancient texts and traditional medicine around the globe. While the anecdotal evidence is substantial, modern scientific evidence for the wind's influence on health is still developing.

The old wives' tale of windy conditions increasing the likelihood of labour has found support in the discovery of a causal relationship between barometric pressure (the pressure within Earth's atmosphere) and birthing. Low barometric pressure induces the rupture of foetal membranes and increases the likelihood of spontaneous delivery.[30] Low barometric, or atmospheric, pressure is associated with high winds – the more slowly the air moves, the higher the pressure in the atmosphere. It is interesting to note that recent research has found the position of the Moon also exerts a significant influence on the timing of labour – an observation once also considered to be a myth.[31]

A 2008 study found many types of weather, including wind, had an influence on human mood.[32] This influence varied between individuals and during different seasons. Increased wind generally had a negative effect on mood, and the negative effect was greater during the warmer seasons of summer and spring than in autumn and winter. The authors of the study suggested that seasonal differences might be due to people spending more time outside in warmer weather but noted that this explanation was highly speculative. A large-scale British research project, dubbed 'Cloudy with a Chance of Pain', found daily fluctuations in weather made a big difference to those experiencing chronic pain. In a survey of more than 13,000 people, it was found that days with higher humidity, lower pressure and stronger winds were more likely to be associated with higher pain.[33]

As well as the wind's general capacity to influence human moods and health, various communities worldwide have observed the power of particular winds to hold distinctive influences. An example of this phenomenon is the foehn wind, found in alpine regions of Switzerland. While the term 'foehn wind' describes a particular weather event in Switzerland, it is also the name of a type of wind. A foehn wind (sometimes known as a chinook) is a wind which is warm, very dry and blows down the leeward side of a mountain. The 'leeward' region of a mountain is the side that faces away from the prevailing wind. The term 'foehn wind' has become a generic one, but its origins lie in the European Alps. In the foehn valleys, the 'foehn wind' has long been colloquially known to cause headaches and a low mood. The name 'foehn' likely originates in the Roman conquest of the territories north of the Alps in the first century BCE. 'Foehn' comes from an Old High German word with roots in the Latin word *favonius*, meaning 'favourable'.[34] Favonius is also the name of a Roman wind deity, the counterpart of the Greek wind god, Zephyrus.

In Alpine folklore, the foehn wind is connected to several health conditions. *Foehnkrankenheit* (foehn sickness) has been thought to involve a range of health conditions including everything from minor headaches and fatigue to insomnia and psychosis. The effect of the foehn wind on mental well-being has seen it linked anecdotally to the incidence of crime waves. When foehn winds blow in Switzerland, the suicide rate is said to increase, and some courts in Swiss cantons have considered the blowing of the foehn winds as an aggravating factor in crime. A 2019 study investigated the potential association of foehn

winds with increased suicide risk in the Tatra mountains that lie on the border of Slovakia and Poland. While the winds did not appear to present a greater daily risk of suicide, they were found to increase the seasonal risk in the area studied, particularly in summer and autumn.[35] Research tying the prevalence of local winds to suicidality highlights the need for further exploration of the connections between human health and the natural environment.

Warm winds from other regions, sharing common features with the foehn winds, are credited with similar effects. In the Dalmatian region of Croatia, the 'ill winds' are the warm southerly *južina* or jugo wind. The winds originate in the Sahara or the Arabian deserts, and like the foehn winds they have a reputation for causing headaches, nausea and poor concentration. An urban myth in Croatia holds that in ancient times, if crimes were committed while the *južina* was blowing, the culprit would be given some leniency.[36] In Israel, the appearance of winds blowing off deserts in the east has been observed to accompany an increase in suicide attempts and psychotic episodes in those suffering mental health issues.[37]

The scientific processes behind the folkloric effects of foehn winds are still developing. In recent years, however, awareness of how regional winds can affect human health, behaviour and well-being has begun to grow. The chinook winds are a regional variety of a foehn-type wind, known for their powerful effect on surrounding landscapes. Chinook winds are often much warmer than the land over which they blow, and these hot, fast-moving winds can vaporize a foot of snow on the ground in just hours. For this reason, chinook winds are called 'snow eaters'. These powerful winds are featured in many myths from First Nations people in Canada and the Rocky Mountains. Salish-speaking tribes have a myth about the great supernatural being Thunderbird driving people from her valley with the Northeast Wind. The movement was a consequence of the devastation in the valley caused by a carelessly lit campfire. Eventually, the myth tells, Thunderbird grew lonely, and told the Northeast Wind to leave, but people would not return while the valley was still cold. The chinook wind was invited to warm the valley, blowing warmth over the land and melting the ice and snow. When life returned to the region, Thunderbird agreed not to let the

197

innocent suffer alongside the careless in the future.[38] Although the winds are presented playing a positive role in this legend, recent research published in the journal *Neurology* showed that migraines could accompany the chinook winds of Alberta, Canada.[39]

Francisco de Goya, *The Snowstorm; or, Winter*, 1786–7, oil on canvas.

Famous winds

As in the Swiss Alps, many areas of the globe are home to named winds which hold distinctive places in their local regions. The act of naming a wind suggests both a reasonable consistency to the quality of the wind and local significance. A named wind will have an identifiable constellation of qualities pertaining to strength, season, direction, temperature and duration.[40] In this way, the name of the wind provides a shorthand for its nature. To express the connection more succinctly, 'a name . . . is a forecast'.[41]

The famed Santa Ana winds of southern California are a type of foehn wind. Most common in the cooler months from October to March, the Santa Anas can also occur from September to June. The winds are known for their powerful blasts, and are considered dangerous due to their ability to rapidly spread wildfires; the Santa Anas are a key element that drives the spread of wildfires in southern California. In the region, the terrain is covered with low-lying vegetation, such as coastal sage brush, and shaped by many canyons. The vegetation-covered canyons create wind tunnels through which the Santa Ana winds can blow, spreading fires – a phenomenon that has increased in severity in recent years.[42]

Winds blow across sand dunes, Fuerteventura, Canary Islands.

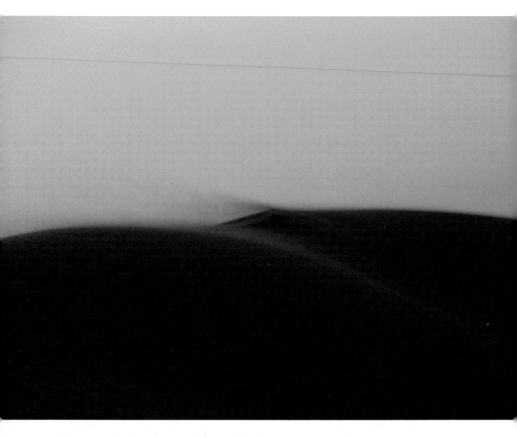

Along with the Santa Anas and chinook winds of the United States, other famous winds known for their unsettling effects can be found across the globe. The sirocco wind of North Africa and the sharav wind in the Middle East are thought to be varieties of foehn winds. Libyans contend with the ghibli, southern France has the mistral wind and the Argentine Andes are home to the zonda. The hot, dry wind that blows off the Sahara affecting parts of North Africa and the Middle East is known as the khamsin. The word *khamsin* in Arabic means 'fifty', and it is thought the wind holds this name as it is said to blow over Egypt in fifty days during spring. The wind has a variety of names in different regions, but one of its uniting features is its capacity to create a sense of unease in the human populations where it travels.

Red Dunes Desert, Dubai.

A welcome wind

The varied effects of windy conditions mean it is perhaps unsurprising to find the negative influence of certain winds can be contrasted against winds considered beneficial to health. The Dutch concept of *uitwaaien* reflects the power of windy weather to improve health and mental outlook. The term *uitwaaien* literally translates as 'out-blowing', conjuring misleading images of candles atop birthday cakes. Instead, the term means 'to walk in the wind'.[43] In Dutch culture, *uitwaaien* involves undertaking physical activity in windy, cool weather. The benefits of the practice are thought to come from replacing 'bad air' with 'good air', to 'blow away' negative feelings and improve energy levels and mental clarity. The practice is thought to be over a hundred years old, with modern scientific studies lending support to the underlying concept that outdoor activity can boost well-being. While there are many health benefits to being outdoors, the influence of different winds on human well-being has been tentatively linked to the ionic charge of air that is created in particular environmental conditions. Ions are molecules that have lost or gained electric charges due to a change in the number of electrons in their make-up. This change from an electrically neutral position can be caused by sunlight, radiation or moving air and water. Air that is heavy with positively charged ions tends to be equated with negative effects on well-being, while the reverse can be found for air with negatively charged ions.

Western Australia is home to a wind known as the 'Fremantle Doctor'. During the summer months, this cooling afternoon sea breeze brings relief from high temperatures. The Doctor's beneficial effects have been recognized in Western Australia since the nineteenth century; it is considered a reliable wind, arriving with consistency and predictable intensity. Assumptions of the Doctor's dependability were subverted during the America's Cup sailing trials of 1986. The arrival of the wind during the trials tipped the balance of fate in favour of Dennis Connor, the skipper of the *Stars and Stripes*, as described in an article in the *Los Angeles Times*:

Waves driven by 28-knot winds knocked boats around like
punching bags, swept over decks and forced people to hang
on for their lives in the America's Cup trials Wednesday –
and that was only in the spectator fleet . . . The Fremantle
Doctor has officially arrived, bringing with him the south-
west summer blasts that turn the Gage Roads channel
between Rottnest Island and the mainland into one of the
world's wildest wind tunnels.[44]

On the other side of the ocean, another Doctor wind is
thought to bring relief in the form of a sea breeze. The Cape
Doctor blows across the western Cape of southern Africa, particularly in Cape Town, from the southeast. Early European
settlers in the region nicknamed the wind the 'Cape Doctor', as
it was thought to have healing properties. The cool breeze of the
Doctor was considered to blow away hot, disease-causing urban
air, sweeping the 'bad' air out to sea.

The stilling

While winds have been blowing for longer than the world has
been spinning, the nature and power of wind has not been constant. Indeed, a significant change in global wind speeds, known
as the 'stilling', has been observed in the last few decades. While
the phenomenon of global stilling is likely to have widespread
consequences, the causes and effects of this change are currently
not well understood. Very recently, scientific measurements of
wind have observed a further change – instead of winds slowing
down on a global scale, winds now appear to be speeding up, in
an intriguing reversal of the earlier trend.[45]

Starting in the 1970s, and continuing in the three decades
that followed, scientists observed a substantial slowing of global
wind speed.[46] While the causes of the decline – considered to
be a drop of between 5 and 15 per cent – remain unknown, the
influence of climate change on global temperatures is thought
to be an important factor in understanding the global patterns
of wind behaviour. Scholars had posited that the Earth's surface

may have been becoming rougher through urban development and increasingly concentrated man-made structures. This roughness was thought to have created a buffer for winds, slowing their movement. Yet while urban development continues, wind speeds in the past two decades have begun to accelerate. For some, this is good news: a continued global stilling was predicted to lead to a drop by as much as 50 per cent in the capacity of the global wind energy industry. Increasing wind speeds will likely lead to greater yields from wind farming technology, at least in the short term.

Increases in the global surface wind speed are problematic for areas affected by wildfire. High-speed winds fuel fires and enhance the rapidity and reach of their spread. High winds were a significant factor in fuelling Australia's unprecedented Black Summer bushfires of 2019–20, and the devastating wildfires in California around the same time. It is currently uncertain whether the noted increase in global wind speeds since 2010 reflects a recovery to 'normal' levels, or forms part of a trend of wind speeds that may result in further increases.

While scholars are currently still debating the direction of the trends in terrestrial wind speeds globally, there is greater consensus around recent patterns of winds involved in major weather events. Meteorological research has shown that in the past four decades a clear upward trend can be observed in the intensity of hurricanes.[47] Similarly, whether the overall trend in global wind speed is becoming faster or slower, evidence is emerging that global wind patterns are shifting in unprecedented ways. Over the last few decades, westerly winds, which traditionally blow westwards across the planet's latitudes, appear to be shifting towards Earth's poles.[48] The change in the distribution and behaviour of global winds holds significant implications for patterns of rainfall, storm systems and ecosystems, although the nature of the effects of this change is yet to be determined.

A further area in which changing wind patterns has already shown an impact is aviation, which has seen a marked increase in clear-air turbulence (less technically known as 'in-flight bumpiness').[49] Clear-air turbulence is the sudden, severe turbulence that occurs in cloudless, or 'clear' skies. It is one of the largest

causes of weather-related aviation incidents; it is difficult for air-craft pilots to avoid, as it is not detectable by on-board sensors and is invisible to the naked eye. Several studies in the past ten years have shown that clear-air turbulence, driven by wind shear, has intensified.[50] It is thought this may be due to climate change, as variations in wind shear are propelled by temperature changes. While temperature differences between the poles and the equa-tor have evened out in recent years in the atmosphere below the stratosphere, temperatures above the stratosphere have cooled, with the resulting imbalance causing an increase in turbulence. A study of upper-level North Atlantic jet streams found a 15 per cent increase in shear in the four decades since 1979.[51] The study showed that climate change was having a greater effect on turbulence than previously had been thought, with climate pro-jections suggesting turbulence at cruising altitudes will continue to increase throughout the world in future years.

Jan Brueghel the Elder, *Allegory of Air*, 1611, oil on canvas.

With the modern technological study of climate and meteorology still relatively new, there is much about wind, its effects and its mechanisms which remains nebulous. Despite the many uncertainties, it is clear that the critical role played by wind makes urgent and important work of better understanding the observed differences in global wind speed.

Wind, with its invisible, ineffable force, has shaped the Earth, the cosmos and the course of human civilization for millennia. While it may be unseen, and often even unnoticed, wind has influenced the evolutionary journeys of humans and animals, and it continues to wield a powerful effect on the behaviours and health of living things. With so many of the planet's processes tightly entwined with wind, it is in our interest to understand the phenomenon more deeply – and to carefully observe global changes in wind speeds and distributions.

By turns destructive and creative, the most constant element in the cultural history of wind is its changeability. At the close of this book, many aspects of wind remain intangible, reflecting a boundless topic with an almost infinite reach through time and space. The cultural history of wind, in all of its complexity, is a continually evolving narrative as essential as it is invisible.

TIMELINE

190 MYA Winds begin to erode the landform known as 'The Wave' in the Navajo sandstone rocks of the Vermilion Cliffs National Monument in Arizona

340 BCE Aristotle writes his influential weather work *Meteorologica*

216 BCE Hannibal leads the Carthaginians to victory over the larger Roman army in the Battle of Cumae, by stationing his troops strategically to benefit from the wind direction

65 CE Roman philosopher Seneca writes *Natural Questions*

1200s Two great winds, known as *kamikaze*, save the Japanese nation from invasion by the Mongols

1300s Windmills become a popular type of power generator throughout Europe

1461 Favourable winds help the Yorkists defeat the Lancastrians in a key battle during the War of the Roses

1540 The first known anemometer invented by the Renaissance artist and cryptographer Leon Battista Alberti in Italy

1590 José de Acosta's *Natural and Moral History of the Indies* is published, giving a detailed study of trade winds

1622 Francis Bacon's *History of the Winds* is published

1672 English physicist Robert Hooke creates the first rotation, or windmill, anemometer

1703 The 'Great Storm' strikes Britain, considered the most powerful to have occurred in the country's recorded history

1889 Vincent van Gogh's *The Starry Night* painted; it depicts the movement of wind with mathematical precision

1934 Australian-British writer P. L. Travers's series *Mary Poppins* opens with the eponymous nanny arriving on the East Wind

1954 Testing of nuclear fusion bombs on the Bikini Atoll by the U.S. military is adversely affected when the winds do not follow the predicted patterns of the meteorologists. As a result, strong westerlies carry fallout contamination across the population of the Marshall Islands, and beyond

1962 American robotic space probe Mariner 2 journeys to Venus and measures the solar wind

1979 Typhoon Tip peaks at over 300 km/h (185 mph), making it the fastest cyclone ever recorded

1996 Australia's Barrow Island records the fastest wind speed ever measured, independent of a tornado. American blockbuster disaster film *Twister* brings storm science to the silver screen

1999 Two cyclones in western and central Europe blow down 176 million m³ (6.2 billion ft³) of wood, valued at €6–7 billion

2005 Hurricane Katrina breaks the levies in New Orleans, USA

2007 The first offshore wind turbine is deployed

2013 'Space winds' or plasmapheric winds are discovered travelling around the edges of our atmosphere at speeds of around 5,000 km/h (3,000 mph)

2015 Archaeologists working in Belize unearth a precious artefact covered in Maya hieroglyphs called a 'wind jewel'; it contains written text and has a vessel likely depicting the Maya wind deity

2018 Winter Olympics powered by wind farms are held in South Korea

2021 Quad-State Tornado strikes the states of Arkansas, Missouri, Tennessee and Kentucky, causing widespread devastation. It is the first tornado in U.S. history to cross through four states

REFERENCES

1 Wind: Natural History

1 Paul Kapp et al., 'Wind Erosion in the Qaidam Basin, Central Asia: Implications for Tectonics, Paleoclimate, and the Source of the Loess Plateau', *GSA Today*, XXI/4–5 (April/May 2011), pp. 4–10.
2 Jan Rocha, 'Drought Bites as Amazon's "Flying Rivers" Dry Up', *The Guardian*, www.theguardian.com, 15 September 2014.
3 Douglas Sheil, 'Trees and Water: Don't Underestimate the Connection', *ZME Science*, www.zmescience.com, 29 October 2019.
4 Fred Pearce, 'A Controversial Russian Theory Claims Forests Don't Just Make Rain – They Make Wind', *Science*, www.science.org, 18 June 2020.
5 Fred Pearce, 'Rivers in the Sky: The Devastating Effect Deforestation Is Having on Global Rainfall', *South China Morning Post*, www.scmp. com, 8 November 2019.
6 Hanna Tuomisto, Matleena Tuomisto and Jouni T. Tuomisto, 'How Scientists Perceive the Evolutionary Origin of Human Traits: Results of a Survey Study', *Ecology and Evolution*, VIII/6 (March 2018), pp. 3518–33.
7 Ian Gilligan, 'Neanderthal Extinction and Modern Human Behaviour: The Role of Climate Change and Clothing', *World Archaeology*, XXXIX/4 (December 2007), pp. 499–514.
8 Ibid., p. 506.
9 M. R. Wright, 'Presocratic Cosmologies', in *The Oxford Handbook of Presocratic Philosophy*, ed. Patricia Curd and Daniel W. Graham, ebook (Oxford, 2008).
10 Aristotle, *Meteorologica*, trans. H.D.P. Lee, ebook (Cambridge, MA, 1952), line 89.
11 V. Coutant and V. Eichenlaub, 'The *De ventis* of Theophrastus: Its Contributions to the Theory of Winds', *Bulletin of the American Meteorological Society*, LV/12 (December 1974), p. 1454.

12 Vitruvius, *The Ten Books on Architecture*, Book 1, Chapter 6, line 2, trans. Morris Hicky Morgan, ebook (Cambridge, MA, 2013).

13 Coutant and Eichenlaub, 'The *De ventis* of Theophrastus', pp. 1454–62.

14 Gareth Williams, 'Seneca on Winds: The Art of Anemology in "Natural Questions" 5', *American Journal of Philology*, CXXVI/3 (Autumn 2005), p. 417.

15 Craig Martin, 'Francis Bacon, José de Acosta, and Traditions of Natural Histories of Winds', *Annals of Science*, LXXVII/4 (October 2020), p. 459.

16 Michael Allaby, *Atmosphere: A Scientific History of Air, Weather, and Climate* (New York, 2009), p. 56.

17 Allison Lee Palmer, *Leonardo da Vinci: A Reference Guide to His Life and Works* (Lanham, MD, 2019), p. 10.

18 V. Obridko and O. Vaisberg, 'On the History of the Solar Wind Discovery', *Solar System Research*, LI/2 (March 2017), pp. 165–9.

19 Eugene Parker, 'Discovering Solar Wind – Eugene Parker's Story', interview by UChicago Creative, 31 May 2017, https://creative. uchicago.edu.

20 Stanislav Boldyrev, Cary Forest and Jan Egedal, 'Electron Temperature of the Solar Wind', *Proceedings of the National Academy of Sciences*, ebook (2020).

21 Stuart Gary, 'Earth Generates Its Own Solar Wind', ABC, www.abc. net.au, 3 July 2013.

22 Elizabeth Howell, '"Winds" from Monster Black Holes Can Rapidly Change Their Temperature', *Space*, www.space.com, 1 March 2017.

23 L. A. Fisk and J. C. Kasper, 'Global Circulation of the Open Magnetic Flux of the Sun', *Astrophysical Journal Letters*, DCCCXCIV/1 (April 2020), pp. 1–5.

24 A. Alabdulgader et al., 'Long-Term Study of Heart Rate Variability Responses to Changes in the Solar and Geomagnetic Environment', *Scientific Reports*, VIII/1 (February 2018), p. 2663.

25 Yuri Zaisev, 'Magnetic Storms Affect Humans as Well as Telecommunications', *Space Daily*, www.spacedaily.com, 29 May 2006.

26 Katelyn N. Allers et al., 'A Measurement of the Wind Speed on a Brown Dwarf', *Science*, CCCLXVIII/6487 (April 2020), pp. 169–72.

27 Helen Phillips, Benoit Legresy and Nathan Bindoff, 'Explainer: How the Antarctic Circumpolar Current Helps Keep Antarctica Frozen', https://theconversation.com, 16 November 2018.

28 Torben Struve et al., 'Middle Holocene Expansion of Pacific Deep Water into the Southern Ocean', *Proceedings of the National Academy of Sciences of the United States of America*, CXVII/2 (January 2020), pp. 889–94.

29 Kapp et al., 'Wind Erosion', pp. 4–10.

30 Ibid., pp. 7–8.
31 Mitchell McMillan et al., 'Large-Scale Cenozoic Wind Erosion in the Puna Plateau: The Salina del Fraile Depression', *Journal of Geophysical Research: Earth Surface*, cxxv/9 (September 2020).
32 Ellen Gray, 'nasa Satellite Reveals How Much Saharan Dust Feeds Amazon's Plants', www.nasa.gov, 23 February 2015.
33 Texas Tech University, 'Giant Pterosaur Needed Cliffs, Downward-Sloping Runways to Taxi, Awkwardly Take Off into Air', *Science Daily*, www.sciencedaily.com, 7 November 2012.
34 Sylvia Hughes, 'Antelope Activate the Acacia's Alarm System', *New Scientist*, www.newscientist.com, 29 September 1990.
35 Barry Gardiner, Peter Berry and Bruno Moulia, 'Review: Wind Impacts on Plant Growth, Mechanics and Damage', *Plant Science*, ccxxv (April 2016), pp. 94–118.
36 A. Miri, D. Dragovich and Z. Dong, 'Vegetation Morphologic and Aerodynamic Characteristics Reduce Aeolian Erosion', *Scientific Reports*, vii/1 (October 2017).
37 Emmanuel de Langre, 'Effects of Wind on Plants', *Annual Review of Fluid Mechanics*, xl/1 (January 2008), p. 142.
38 Damien Carrington, 'uk Faces Sharp Rise in Wind Storms and Higher Bills as World Warms', *The Guardian*, www.theguardian.com, 16 May 2017.
39 Patrick Barkham, 'Bees May Struggle in Winds Caused by Global Warming, Study Finds', *The Guardian*, www.theguardian.com, 18 February 2020.
40 Thomas A. Clay et al., 'Sex-Specific Effects of Wind on the Flight Decisions of a Sexually Dimorphic Soaring Bird', *Journal of Animal Ecology*, lxxxix/9 (August 2020), pp. 1–13.
41 Langre, 'Effects of Wind', p. 111.
42 Ibid.

2 Wind in the World's Oldest Literature

1 Thorkild Jacobsen, 'The Líl of dEn-líl', in *dumu-é-dub-ba-a: Studies in Honor of Åke W. Sjöberg*, ed. Hermann Behrens, Darlene Loding and Martha T. Roth (Philadelphia, pa, 1989), p. 269.
2 See Louise M. Pryke, *Ishtar: Gods and Heroes of the Ancient World* (London and New York, 2017).
3 M. Hutter, 'Lillith', in *Dictionary of Deities and Demons in the Bible: ddd*, ed. Pieter Willem van der Horst, Bob Becking and K. van der. Toorn (Grand Rapids, mi, 1999), p. 520.
4 JoAnn Scurlock, 'Soul Emplacements in Ancient Mesopotamian Funerary Rituals', in *Magic and Divination in the Ancient World*, ed. Leda Jean Ciraolo and Jonathan Lee Seidel (Leiden, 2002), pp. 1–6.

5 Proverbs: Collection 4, ETCSL t.6.1.04, in J. A. Black et al., *The Electronic Text Corpus of Sumerian Literature*, 1998–2006, available at http://etcsl.orinst.ox.ac.uk.

6 Shlomo Izre'el, *Adapa and the South Wind: Language Has the Power of Life and Death* (Winona Lake, IN, 2001), pp. 16–19.

7 Charles G. Leland, *The Algonquin Legends of New England; or, Myths and Folk Lore of the Micmac, Passamaquoddy, and Penobscot Tribes*, ebook (London, 1884).

8 Ibid.

9 Ibid.

10 Hugo Winckler, *Der Thontafelfund von el Amarna* (Berlin, 1890).

11 D. R. Jackson, 'Solomon', in *Dictionary of the Old Testament: Wisdom, Poetry and Writings*, ed. Tremper Longman III and Peter Enns (Nottingham, 2008), p. 736.

12 Ibid.

13 *The Epic of Gilgamesh*, Standard Babylonian Version II, in *The Standard Babylonian Gilgamesh Epic*, trans. Andrew George (Oxford, 2003), pp. 234–5.

14 Mohammed Elsayed, 'Remarks on the Concept of the Role of Wind in the Texts of the Temple of Esna', *Shedet*, V/5 (December 2018), pp. 82–95.

15 Maulana Karenga, *Maat, the Moral Ideal in Ancient Egypt: A Study in Classical African Ethics* (London, 2012), p. 37.

16 Foy Scalf, ed., *Book of the Dead: Becoming God in Ancient Egypt* (Chicago, IL, 2017), p. 236.

17 Cecilia Grave, 'Northwest Semitic Ṣapānu in a Break-Up of an Egyptian Stereotype Phrase in EA 147', *Orientalia*, LI/2 (1982), p. 171.

18 Scott B. Noegel, 'On the Wings of the Winds: Towards an Understanding of Winged Mischwesen in the Ancient Near East', *KASKAL: Rivista di Storia, ambiente e culture del Vicino Oriente antico*, XIV (2017), p. 19.

3 Myth, Folklore and Religion

1 Daniel Merkur, 'Breath-Soul and Wind Owner: The Many and the One in Inuit Religion', *American Indian Quarterly*, VII/3 (Summer 1983), pp. 23–39.

2 Alex Purves, 'Wind and Time in Homeric Epic', *Transactions of the American Philological Association*, CXL/2 (Autumn 2010), pp. 323–50.

3 Ibid., p. 333.

4 Thomas L. Hankins and Robert J. Silverman, *Instruments and the Imagination* (Princeton, NJ, 2014), p. 87.

5 Samuel Coleridge, 'The Eolian Harp', in *Sibylline Leaves* (London, 1817), p. 177.

6   M. H. Abrams, "The Correspondent Breeze: A Romantic
     Metaphor', *Kenyon Review*, xix/1 (Winter 1957), p. 114.
7   Virgil, 'Georgics 3.267', in *Eclogues, Georgics, Aeneid: Books 1-6*, trans.
     H. Rushton Fairclough and Rev. G. P. Goold, ebook (Cambridge,
     MA, 1999).
8   Aristotle, *The History of Animals*, vi.2, trans. D'Arcy Wentworth
     Thompson, ebook (Oxford, 1910).
9   Conway Zirkle, 'Animals Impregnated by the Wind', *Isis*, xxv/1 (May
     1936), pp. 95–130.
10 M. David Litwa, 'Divine Conception', in *How the Gospels Became
     History: Jesus and Mediterranean Myths* (London and New Haven, CT,
     2019), p. 91.
11 Jacob Grimm, *Teutonic Mythology*, trans. James Steven Stallybrass
     (London, 1880), p. 918.
12 Tamra Andrews, *Legends of the Earth, Sea, and Sky: An Encyclopedia of
     Nature Myths* (Santa Barbara, CA, 1998), p. 256.
13 C. Prager and G. Braswell, 'Maya Politics and Ritual: An Important
     New Hieroglyphic Text on Carved Jade from Belize', *Ancient
     Mesoamerica*, xxvii/2 (Fall 2016), pp. 267–78.
14 Karl A. Taube, 'The Symbolism of Jade in Classic Maya Religion',
     Ancient Mesoamerica, xvi/1 (January 2005), pp. 23–50.
15 Lucia R. Henderson, 'Earflare Set 3rd–9th Century', www.
     metmuseum.org, accessed 31 October 2020.
16 Taube, 'Symbolism', p. 31.
17 Prager and Braswell, 'Maya Politics and Ritual', p. 276.
18 Andrews, *Legends*, pp. 189–90.
19 R. Flikke, 'Breathing Pneumatology: Spirit, Wind, and Atmosphere
     in a Zulu Zionist Congregation', in *Faith in African Lived
     Christianity: Bridging Anthropological and Theological Perspectives*,
     ed. K. Lauterbach and M. Vähäkangas (Boston, MA, 2020), p. 299.
20 Ibid.
21 Dana E. Powell, *Landscapes of Power: Politics of Energy in the Navajo
     Nation* (Durham, NC, 2018), p. 267.
22 Douglas Duckworth, 'Rangjung Dorjé's Key to the Essential
     Points of Wind and Mind', *Buddhism and Medicine: An Anthology of
     Premodern Sources*, ed. C. Pierce Salguero (New York, 2017), p. 413.
23 John Bierhorst, 'In the Trail of the Wind', in *Sing with the Heart of a
     Bear: Fusions of Native and American Poetry, 1890–1999*, ed. Kenneth
     Lincoln (Berkeley, CA, 2000), p. 308.
24 Lawrence Eugene Sullivan, *Native Religions and Cultures of North
     America: Anthropology of the Sacred* (London, 2003), p. 20.
25 Lewis Henry Morgan, *The League of the Iroquois* (North Dighton, MA,
     1999), p. 141.
26 Andrews, *Legends*, pp. 236–7.

27 Ibid., p. 236.
28 Robert D. Craig, *Handbook of Polynesian Mythology* (Santa Barbara, CA, 2004), p. 194.
29 Martha Warren Beckwith, *Hawaiian Mythology* (Honolulu, HI, 1971), p. 86.

4 Warring Winds

1 Stephen Moss, 'Weatherwatch: 1703 Storm Was First Media Weather Event of Modern Age', *The Guardian*, www.theguardian.com, 10 November 2018.
2 Timothy J. Jorgensen, 'Bikini Islanders Still Deal with Fallout of U.S. Nuclear Tests, More than 70 Years Later', https://theconversation.com, 29 June 2016.
3 Livy, *The History of Rome*, XXII.43, trans. Rev. Canon Robert (New York, 1912), pp. 10–11.
4 Lucretius, *On the Nature of Things*, Book 5, trans. W.H.D. Rouse and Martin F. Smith (Cambridge, MA, 1924), p. 750; Columella, *On Agriculture*, Book 5, trans. E. S. Forster and Edward H. Heffner (Cambridge, MA, 1954) p. 15.
5 Trevor Absolon, *Samurai Armour*, vol. I: *The Japanese Cuirass* (New York, 2017), p. 170.
6 Devin Powell, 'Japan's Kamikaze Winds, the Stuff of Legend, May Have Been Real', *National Geographic*, www.nationalgeographic.com, 5 November 2014.
7 J. D. Woodruff et al., 'Depositional Evidence for the Kamikaze Typhoons and Links to Changes in Typhoon Climatology', *Geology*, XLIII/1 (January 2015), pp. 91–4.
8 Ibid.
9 George Goodwin, *Fatal Colours: The Battle of Towton, 1461*, ebook (London, 2012).
10 Ibid.
11 Ibid.
12 I.A.A. Thompson, 'The Appointment of the Duke of Medina Sidonia to the Command of the Spanish Armada', *Historical Journal*, XII/2 (June 1969), pp. 212–13.
13 Ron Chernow, *Washington: A Life* (New York, 2010), p. 246.
14 Gillian Brockell, 'Weird Weather Saved America Three Times', *Washington Post*, www.washingtonpost.com, 20 March 2019.
15 Gerald J. Fitzgerald, 'Chemical Warfare and Medical Response during World War I', *American Journal of Public Health*, XCVIII/4 (July 2008), pp. 611–25.
16 Bob Yirka, 'Trace Amounts of Isotope from Fukushima Disaster Found in California Wine', https://phys.org, 24 July 2018.

17 Jonathan Bate, 'Living with the Weather', *Studies in Romanticism*, xxxv/3 (Fall 1996), p. 439.
18 Jan Golinski, *British Weather and the Climate of Enlightenment* (Chicago, IL, 2011), p. 215.
19 John Withington, *Storm* (London, 2016), p. 21.
20 Kerry A. Emanuel, *Divine Wind: The History and Science of Hurricanes* (New York, 2005), p. 18.
21 Ibid.
22 Dennis Wheeler and Clive Wilkinson, 'From Calm to Storm: The Origins of the Beaufort Wind Scale', *Mariner's Mirror*, xc/2 (2004), p. 189.
23 Scott Huler, *Defining the Wind: The Beaufort Scale, and How a Nineteenth-Century Admiral Turned Science into Poetry* (New York, 2004), p. 85.
24 Moss, 'Weatherwatch'.
25 Daniel Defoe, *The Storm; or, a Collection of the Most Remarkable Casualties and Disasters which Happen'd in the Late Dreadful Tempest, Both by Sea and Land*, ebook (London, 2013).
26 Huler, *Defining the Wind*, p. 22.
27 Meaghan Evans, 'Earth's Strongest, Most Massive Storm Ever', *Scientific American*, www.scientificamerican.com, 12 October 2012.
28 Kerry Emanuel, 'Climate Change and Hurricane Katrina: What Have We Learned?', https://theconversation.com, 24 August 2015.
29 Bernard Mergen, *Weather Matters: An American Cultural History since 1900* (Lawrence, KS, 2008), p. 281.
30 William K. Stevens, 'Tetsuya Fujita, 78, Inventor of Tornado Scale', *New York Times*, www.nytimes.com, 21 November 1998.

5 Trade and Technology

1 Paul-Alain Beaulieu, *A History of Babylon, 2200 BC–AD 75*, ebook (Hoboken, NJ, 2018).
2 Robert A. Carter, 'Watercraft of the Ancient Near East', in *A Companion to the Archaeology of the Ancient Near East*, ed. D. T. Potts (Malden, MA, 2012), pp. 349–50.
3 Ibid., p. 349.
4 K. Kris Hirst, 'Mesopotamian Reed Boats Changed the Stone Age', www.thoughtco.com, 12 November 2019.
5 Lewis Dartnell, *Origins: How the Earth Made Us*, ebook (London, 2019).
6 Rafe de Crespigny, *A Biographical Dictionary of Later Han to the Three Kingdoms (23–220 AD)* (Leiden, 2007), p. 184.
7 Cesare Rossi and Flavio Russo, *Ancient Engineers' Inventions: Precursors of the Present* (Cham, 2017), p. 348.

8 John Noble Wilford, 'Ancient Smelter Used Wind to Make High-Grade Steel', *New York Times*, www.nytimes.com, 6 February 1996.

9 Ibid.

10 Ben Coates, *Why the Dutch are Different: A Journey into the Hidden Heart of the Netherlands*, ebook (London, 2015).

11 Kris De Decker, 'Wind Powered Factories: History (and Future) of Industrial Windmills', *Low-Tech Magazine*, www.lowtechmagazine.com, 8 October 2010.

12 National Museum of Australia, 'Defining Moments: Earliest Evidence of the Boomerang in Australia', National Museum of Australia, www.nma.gov.au, accessed 10 November 2020.

13 Pawel Valde-Nowak, Adam Nadachowski and Mieczyslaw Wolsan, 'Upper Palaeolithic Boomerang Made of a Mammoth Tusk in South Poland', *Nature*, cccxxix/6138 (October 1987), p. 437.

14 Michael Westaway et al., 'The Death of Kaakutja: A Case of Peri-Mortem Weapon Trauma in an Aboriginal Man from North-Western New South Wales, Australia', *Antiquity*, xc/353 (October 2016), pp. 1318–33.

15 '"Talibanned:" Favourite Afghan Pastimes Again under Threat', *Al Jazeera*, www.aljazeera.com, 8 July 2021.

16 Bruno Chanetz, 'A Century of Wind Tunnels since Eiffel', *Comptes Rendus Mécanique*, cccxlv/8 (August 2017), pp. 581–94.

17 Robert W. Righter, *Wind Energy in America: A History* (Norman, ok, and London, 1996), p. 42.

18 Michael L. Ross, *The Oil Curse: How Petroleum Wealth Shapes the Development of Nations* (Princeton, nj, 2012), p. 52.

19 Ibid.

20 Susan Gourvenec, 'Floating Wind Farms: How to Make Them the Future of Green Electricity', https://theconversation.com, 20 July 2020.

21 Charles Choi, 'Out-of-this-World Proposal for Solar Wind Power', *New Scientist*, www.newscientist.com, 24 September 2010.

22 Brooks L. Harrop and Dirk Schulze-Makuch, 'The Detection of a Dyson-Harrop Satellite: A Technologically Feasible Astroengineering Project and Alternative to the Traditional Dyson Sphere Using the Solar Wind', *International Journal of Astrobiology*, ix (2010), pp. 89–99.

23 Tanja M. Straka, Marcus Fritze and Christian C. Voigt, 'The Human Dimensions of a Green–Green-Dilemma: Lessons Learned from the Wind Energy–Wildlife Conflict in Germany', *Energy Reports*, vi (November 2020), pp. 1768–77.

24 Benjamin K. Sovacool, 'Contextualizing Avian Mortality: A Preliminary Appraisal of Bird and Bat Fatalities from Wind,

Fossil-Fuel, and Nuclear Electricity', *Energy Policy*, xxxvii/6 (June 2009), pp. 2241–8.

25   Simon Chapman, 'Wind Farms Are Hardly the Bird Slayers They're Made Out to Be. Here's Why', https://theconversation.com, 16 June 2017.

26   Mathew Denholm, 'Paint It Black: Wind Farm Commissioner Andrew Dyer Backs Calls for Coloured Turbines to Save Birds', *The Australian*, www.theaustralian.com.au, 30 August 2020.

27   Roel May et al., 'Paint It Black: Efficacy of Increased Wind Turbine Rotor Blade Visibility to Reduce Avian Fatalities', *Ecology and Evolution*, x/16 (August 2020), pp. 8927–35.

28   D. S. Mǎntoiu et al., 'Wildlife and Infrastructure: Impact of Wind Turbines on Bats in the Black Sea Coast Region', *European Journal of Wildlife Research*, lxvi/3 (June 2020), pp. 1–13.

29   Panu Maijala et al., 'Infrasound Does Not Explain Symptoms Related to Wind Turbines', *Publications of the Government's Analysis, Assessment and Research Activities*, 34 (June 2020), pp. 1–169.

30   F. Hoffmann, 'Ancient Egypt', in *The Cambridge History of Magic and Witchcraft in the West: From Antiquity to the Present*, ed. S.J.D. Collins (Cambridge, 2015), p. 77.

31   Mark Edward Lewis, *Sanctioned Violence in Early China* (New York, 1990), p. 219.

32   Sam Kean, 'The Chemist Who Thought He Could Harness Hurricanes', *The Atlantic*, www.theatlantic.com, 5 September 2017.

33   James Rodger Fleming, *Fixing the Sky: The Checkered History of Weather and Climate Control* (New York, 2010), p. 151.

34   Umair Irfan, 'No, You Can't Just Nuke a Hurricane. But There Are Other Options', www.vox.com, 26 August 2019.

35   Eleanor Cummins, 'With Operation Popeye, the u.s. Government Made Weather an Instrument of War', *Popular Science*, www.popsci.com, 20 March 2018.

36   Seymour M. Hersh, 'Rainmaking Is Used as Weapon by u.s', *New York Times*, www.nytimes.com, 3 July 1972; Jack Anderson, 'Air Force Has Its Rainmakers Imitating Monsoons', *Washington Post*, www.washingtonpost.com, 18 March 1971.

37   Thomas A. Easton, *Taking Sides: Clashing Views in Energy and Society* (New York, 2009), p. 230.

6 Art, Literature and Popular Culture

1   Thomas Higham et al., 'Testing Models for the Beginnings of the Aurignacian and the Advent of Figurative Art and Music: The Radiocarbon Chronology of Geißenklösterle', *Journal of Human Evolution*, lxii/6 (June 2012), pp. 664–76.

2 John Branch, 'The Most Dominant Force at the Olympics? Wind', *New York Times*, www.nytimes.com, 15 February 2018.

3 Norman Joseph William Thrower, *Maps and Civilization: Cartography in Culture and Society* (Chicago, IL, 2007), p. 187.

4 J. L. Aragón et al., 'Turbulent Luminance in Impassioned van Gogh Paintings', *Journal of Mathematical Imaging and Vision*, XXX/3 (March 2008), pp. 275–83.

5 Ibid.

6 Sarah E. Earp and Donna L. Maney, 'Birdsong: Is It Music to Their Ears?', *Frontiers in Evolutionary Neuroscience*, IV/14 (November 2012), pp. 1–10.

7 Higham et al., 'Testing Models', pp. 664–76.

8 BBC News, 'Earliest Music Instruments Found', www.bbc.com, 25 May 2012.

9 Jack Guy, 'Listen to the Sound of an 18,000-Year-Old Musical Instrument', https://edition.cnn.com, 11 February 2020.

10 Stevie Chick, 'Wind of Change: Did the CIA Write the Cold War's Biggest Anthem?', *The Guardian*, www.theguardian.com, 15 May 2020.

11 Mark Everard, *Breathing Space: The Natural and Unnatural History of Air*, ebook (London, 2015).

12 Ibid.

13 Jack Whatley, 'The Story Behind the Song: "Blowin' in the Wind", the Bob Dylan Classic Written in 10 Minutes', Far Out, https://faroutmagazine.co.uk, 9 July 2020.

14 Seth Rogovoy, *Bob Dylan: Prophet, Mystic, Poet* (New York, 2009), pp. 65–6.

15 'The West Wind; Meteorology and Myth', *The Economist*, www.economist.com, 29 December 2017.

16 Robert Louis Stevenson, *A Child's Garden of Verse* (London, 1895), p. 46.

17 Arthur Conan Doyle, 'His Last Bow', ebook (London, 2000).

18 Andrew Glazzard, 'The East Wind', in *The Case of Sherlock Holmes: Secrets and Lies in Conan Doyle's Detective Fiction* (Edinburgh, 2018), pp. 179–92.

19 P. L. Travers, *Mary Poppins* (London, 2008), p. 20.

20 Brian Szumsky, '"All That Is Solid Melts into the Air": The Winds of Change and Other Analogues of Colonialism in Disney's Mary Poppins', *The Lion and the Unicorn*, XXIV/1 (January 2000), p. 105.

21 Charles Dickens, *Bleak House*, ebook (London, 1997).

22 Alfred Stille, 'Acute Articular Rheumatism' in *Medical Record*, XV, ed. George Frederick Shrady and Thomas Lathrop Stedman (New York, 1879), p. 49.

23 Raymond Chandler, 'The Red Wind', in *Trouble Is My Business, and Other Stories* (Harmondsworth, 1946), pp. 69–124.

24 John Needham, 'The Devil Winds Made Me Do It: Santa Anas Are Enough to Make Anyone's Hair Stand on End', *Los Angeles Times*, www.latimes.com, 12 March 1988; Bad Religion, 'Los Angeles Is on Fire', *The Empire Strikes First* (Sound City, LA, 2004).

25 Joan Didion, 'Los Angeles Notebook', in *Slouching towards Bethlehem* (London, 1969), p. 132.

26 Emily Dickinson, 'xxiv. The Wind', in *Poems by Emily Dickinson*, ed. Mabel Loomis Todd and T. W. Higginson (Boston, MA, 1891), p. 96.

27 Steve Connor, 'Why Calls of the Wild are the Secret of a Good Horror Film', *The Independent*, www.independent.co.uk, 26 May 2010.

28 L. Frank Baum, *The Wonderful Wizard of Oz* (Chicago, IL, 1900), pp. 14–15.

29 Bernard Weinraub, 'Film: Hollywood's Newest Boys of Summer', *New York Times*, www.nytimes.com, 12 May 1996.

30 Marshall Shepherd, 'The Death of Bill Paxton Reminds Us that "Twister" Changed Meteorology', *Forbes*, www.forbes.com, 26 February 2017.

31 John Knox, 'Recent and Future Trends in U.S. Undergraduate Meteorology Enrollments, Degree Recipients, and Employment Opportunities', *Bulletin of the American Meteorological Society*, LXXXIX/6 (June 2008), p. 873.

32 Ibid.

33 Shepherd, 'Death of Bill Paxton'.

34 NOAA News, 'NOAA Tornado Scientists Inspired "Twister" Creators 20 Years Ago', www.noaa.gov, 17 June 2016.

35 BBC News, 'Fish Rain Down on Sri Lanka Village', www.bbc.com, 6 May 2014.

36 Jason Samenow, 'December Tornadoes Aren't Rare, but Friday's Outbreak Was Something Totally Different', *Washington Post*, www. washingtonpost.com, 12 December 2021.

37 Adam Poulisse, '"Sharknado:" Can It Happen? Science Says Maybe', *Los Angeles Daily News*, www.dailynews.com, 28 August 2017.

38 Mike Bender, *A New History of Yachting* (Woodbridge, 2017), p. 12.

39 Ibid., p. 18.

40 Ibid., p. 1.

41 Mathew Taylor, 'Serious Money at Stake for 100m Victor', *The Guardian*, www.theguardian.com, 6 August 2012.

42 Conner Hazelrigg, Bryson Waibel and Blane Baker, 'Modeling of Women's 100-m Dash World Record: Wind-Aided or Not?', *Physics Teacher*, LIII/8 (November 2015), pp. 458–60.

43 Chris Cooper, *Run, Swim, Throw, Cheat: The Science Behind Drugs in Sport* (Oxford, 2012), p. 32.

44 Branch, 'The Most Dominant Force at the Olympics? Wind'.

45 Aamer Madhani and Nancy Armour, 'How the Fierce Winds Are Impacting the 2018 Winter Olympics Competition', *USA Today*, www.usatoday.com, 12 February 2018.

7 Wind, the Environment and the Future

1 María Paula Rubiano A., 'These Masters of the Sky Can Fly for Hours (or Days) While Barely Flapping', *Audobon*, www.audobon.org, 4 September 2020.

2 Jorn A. Cheney et al., 'Bird Wings Act as a Suspension System that Rejects Gusts', *Proceedings of the Royal Society B: Biological Sciences*, CCLXXXVII/1937 (October 2020), available at https://royalsocietypublishing.org.

3 Norman Elkins, *Weather and Bird Behaviour* (London, 2004), p. 207.

4 Emily Shepard et al., 'Wind Prevents Cliff-Breeding Birds from Accessing Nests through Loss of Flight Control', *eLife*, 8 (June 2019), available at https://elifesciences.org.

5 Alexander Piel and Fiona Stewart, 'Chimpanzee "Nests" Shed Light on the Origins of Humanity', https://theconversation.com, 26 July 2018.

6 Ibid.

7 Colin M. Donihue et al., 'Hurricane Effects on Neotropical Lizards Span Geographic and Phylogenetic Scales', *Proceedings of the National Academy of Sciences of the United States of America*, CXVII/19 (April 2020), pp. 10429–34.

8 Joshua Sokol, 'Hurricanes Are Reshaping Evolution across the Caribbean', *New York Times*, www.nytimes.com, 27 April 2020.

9 Michael J. Cherry and Brandon T. Barton, 'Effects of Wind on Predator–Prey Interactions', *Food Webs*, XIII (December 2017), pp. 92–7.

10 Ashley R. Hayes and Nancy J. Huntly, 'Effects of Wind on the Behavior and Call Transmission of Pikas (*Ochotona princeps*)', *Journal of Mammalogy*, LXXXVI/5 (October 2005), pp. 974–81.

11 D. F. Ganihar et al., 'Wind-Evoked Evasive Responses in Flying Cockroaches', *Journal of Comparative Physiology A: Sensory, Neural, and Behavioral Physiology*, CLXXV/1 (August 1994), pp. 49–65.

12 Cherry and Barton, 'Effects of Wind'.

13 Y. Choh and J. Takabayashi, 'Predator Avoidance in Phytophagous Mites: Response to Present Danger Depends on Alternative Host Quality', *Oecologia*, CLI/2 (March 2007), pp. 262–7.

14 S. Cooper, 'Optimal Hunting Group Size: The Need for Lions to Defend Their Kills against Loss to Spotted Hyaenas', *African Journal of Ecology*, XXIX/2 (June 1991), p. 133.

15 Cherry and Barton, 'Effects of Wind', p. 96.

16 Annie Sneed, 'Wind Turbines Can Act Like Apex Predators',
    Scientific American, www.scientificamerican.com, 14 November 2018.
17 Amanda Schmidt, 'Researchers' Theory Examines How Weather
    May Hold Clues into the Likelihood of Shark Attacks',
    www.accuweather.com, 3 August 2019.
18 Neil Hammerschlag, R. Aidan Martin and Chris Fallows,
    'Effects of Environmental Conditions on Predator–Prey Interactions
    between White Sharks (*Carcharodon carcharias*) and Cape Fur
    Seals (*Arctocephalus pusillus pusillus*) at Seal Island, South Africa',
    Environmental Biology of Fishes, LXXVI/2–4 (August 2006), pp. 341–50.
19 Ibid.
20 Chris Fallows et al., 'White Sharks (*Carcharodon carcharias*)
    Scavenging on Whales and Its Potential Role in Further Shaping the
    Ecology of an Apex Predator', *PLOS ONE*, VIII/4 (April 2013), available
    at https://journals.plos.org, accessed September 2020.
21 Sarah Strauss, 'An Ill Wind: The Foehn in Leukerbad and Beyond',
    Journal of the Royal Anthropological Institute, XIII/1 (2007),
    pp. S165–S181.
22 Mehrab Dashtdar et al., 'The Concept of Wind in Traditional
    Chinese Medicine', *Journal of Pharmacopuncture*, XIX/4 (December
    2016), pp. 293–302; Chris Low, 'Khoisan Wind: Hunting and
    Healing', *Journal of the Royal Anthropological Institute*, XIII/1 (April
    2007), pp. S71–S90.
23 Hyein Shim, Maria Kim and Doojin Ryu, 'Effects of Intraday
    Weather Changes on Asset Returns and Volatilities', *Zbornik Radova
    Ekonomski Fakultet u Rijeka*, XXXV/2 (December 2017), pp. 301–30.
24 Richard L. Peterson, *Trading on Sentiment: The Power of Minds over
    Markets* (Hoboken, NJ, 2016), p. 73.
25 Hui-Chu Shu and Mao-Wei Hung, 'Effect of Wind on Stock
    Market Returns: Evidence from European Markets', *Applied
    Financial Economics*, XIX/11 (2009), pp. 893–904.
26 Stephen P. Keef and Melvin L. Roush, 'The Weather and Stock
    Returns in New Zealand', *Quarterly Journal of Business and Economics*,
    XLI/1–2 (Winter–Spring 2002), p. 61.
27 Ellen G. Cohn, 'Weather and Crime', *British Journal of Criminology*,
    XXX/1 (Winter 1990), pp. 51–64.
28 Sherry Towers et al., 'Factors Influencing Temporal Patterns in Crime
    in a Large American City: A Predictive Analytics Perspective', *PLOS
    ONE*, XIII/10 (October 2018), available at https://journals.plos.org.
29 Daniella Miletic, 'No One Breezes through Life When the Wind Just
    Blows and Blows', *The Age*, www.theage.com.au, 7 August 2018.
30 O. Akutagawa, H. Nishi and K. Isaka, 'Spontaneous Delivery Is
    Related to Barometric Pressure', *Archives of Gynecology and Obstetrics*,
    CCLXXV/4 (April 2007), pp. 249–54.

31 G. Ugolini, 'Does Lunar Position Influence the Time of Delivery? A Statistical Analysis', *European Journal of Obstetrics, Gynecology, and Reproductive Biology*, LXXVII/1 (March 1998), pp. 47–50.

32 Jaap J. A. Denissen et al., 'The Effects of Weather on Daily Mood: A Multilevel Approach', *Emotion*, VIII/5 (November 2008), pp. 662–7.

33 Jeremy Deaton, 'No Rain, No Pain: Study Links Foul Weather with Body Aches', *Washington Post*, www.washingtonpost.com, 28 May 2020.

34 Hans Richner and Patrick Hächler, 'Understanding and Forecasting Alpine Foehn', in *Mountain Weather Research and Forecasting Recent Progress and Current Challenges*, ed. Fotini K. Chow et al. (Dordrecht, 2013), p. 253.

35 I. Koszewska et al., 'Foehn Wind as a Seasonal Suicide Risk Factor in a Mountain Region', *Psychiatria i Psychologia Kliniczna*, XIX/1 (February 2019), pp. 48–53.

36 Iva Ralika, 'The *Južina* Wind Phenomenon', www.croatiaweek.com, 28 February 2020.

37 Naomy S. Yackerson et al., 'The Influence of Air-Suspended Particulate Concentration on the Incidence of Suicide Attempts and Exacerbation of Schizophrenia', *International Journal of Biometeorology*, LVIII/1 (January 2013), pp. 61–7.

38 Bon I. Whealdon et al., *'I Will Be Meat for My Salish': The Buffalo and the Montana Writers Project Interviews on the Flathead Indian Reservation* (Helena, MT, 2001) pp. 163–5.

39 J. Cooke, S. Rose and J. Becker, 'Chinook Winds and Migraine Headache', *Neurology*, LIV/2 (January 2000), pp. 302–7.

40 Mark Vanhoenacker, 'The Wind Cries . . . Oe?', *New York Times*, www.nytimes.com, 23 December 2013.

41 Ibid.

42 Judy Abel, 'Two Fire Experts Give a 100-Year Local Fire History', *Malibu Times*, https://malibutimes.com, 16 March 2020.

43 Melissa Alvarez, *The Simplicity of Cozy: Hygge, Lagom and the Energy of Everyday Pleasures*, ebook (Woodbury, MN, 2018).

44 Rich Roberts, 'America's Cup Trials: The Fremantle Doctor, a Very Ill Wind, Blows Connor a Lot of Good', *Los Angeles Times*, www.latimes.com, 4 December 1986.

45 Z. Zeng et al., 'A Reversal in Global Terrestrial Stilling and Its Implications for Wind Energy Production', *Nature Climate Change*, 9 (November 2019), pp. 979–85.

46 Tim R. McVicar et al., 'Less Bluster Ahead? Ecohydrological Implications of Global Trends of Terrestrial Near-Surface Wind Speeds', *Ecohydrology*, V/4 (July 2012), pp. 381–8.

47 Kerry Emanuel, 'Evidence that Hurricanes Are Getting Stronger', *PNAS*, CXVII/24 (May 2020), pp. 13194–5.

48 J. T. Abell et al., 'Poleward and Weakened Westerlies during Pliocene Warmth', *Nature*, 589 (January 2021), pp. 70–75.

49 Luke N. Storer, Paul D. Williams and Manoj M. Joshi, 'Global Response of Clear-Air Turbulence to Climate Change', *Geophysical Research Letters*, xliv/19 (October 2017), pp. 9976–84.

50 Ibid.

51 S. H. Lee, P. D. Williams and T.H.A. Frame, 'Increased Shear in the North Atlantic Upper-Level Jet Stream over the Past Four Decades', *Nature*, 572 (August 2019), pp. 639–42.

SELECT BIBLIOGRAPHY

DeBlieu, Jan, *Wind: How the Flow of Air Has Shaped Life, Myth, and the Land* (Berkeley, CA, 2015)

Emanuel, Kerry, *Divine Wind: The History and Science of Hurricanes* (New York, 2005)

Hargrove, Brantley, *The Man Who Caught the Storm: The Life of Legendary Tornado Chaser Tim Samaras* (New York, 2019)

Huler, Scott, *Defining the Wind: The Beaufort Scale and How a Nineteenth-Century Admiral Turned Science into Poetry* (New York, 2004)

Streever, Bill, *Soon I Heard a Roaring Wind: A Natural History of Moving Air* (New York, 2016)

Watson, Lyall, *Heaven's Breath: A Natural History of the Wind* (New York, 1984)

Weidensaul, Scott, *Living on the Wind: Across the Hemisphere with Migratory Birds* (New York, 2000)

Withington, John, *Storm* (London, 2016)

ASSOCIATIONS AND WEBSITES

National Oceanic and Atmospheric Administration (NOAA)
www.noaa.gov

World Climate Research Programme (WCRP)
www.wcrp-climate.org

Societies

American Meteorological Society
www.ametsoc.org/AMS

Australian Meteorological and Oceanographic Society (AMOS)
www.wcrp-climate.org

The Royal Meteorological Society
www.rmets.org/

World Meteorological Organization
https://public.wmo.int/en

Other Wind Information

Real-time plot of solar wind activity
www.swpc.noaa.gov/products/ace-real-time-solar-wind

The NOAA National Severe Storms Laboratory
www.nssl.noaa.gov

Archived information on Operation Popeye: Transcript of the U.S.
Senate Hearing on Weather Modification of 20 March 1974.
https://web.archive.org/web/20090612231729/http://www.sunshine-
project.org/enmod/popeye

ACKNOWLEDGEMENTS

In the process of writing *Wind*, I've been fortunate to have the generous support of colleagues, students, family and friends. My thanks go especially to Samantha Grosser for her enthusiasm and knowledge, and Emma Barlow for helping to navigate Latin texts. I am grateful to the fantastic team at Reaktion Books, particularly Michael Leaman, Daniel Allen and Alex Ciobanu. All mistakes are my own.

PHOTO ACKNOWLEDGEMENTS

The author and publishers wish to express their thanks to the below sources of illustrative material and/or permission to reproduce it. Some locations of artworks are also given below, in the interest of brevity:

From Cleveland Abbe, *The Aims and Methods of Meteorological Work* (Baltimore, MD, 1899), photo NOAA/Sean Linehan, NOS, NGS: p. 22; from Francis Bacon and William Rawley, *Sylva Sylvarum; or, A Natural History in Ten Centuries* (London, 1683), photo Library of Congress, Rare Book and Special Collections Division, Washington, DC: p. 18; Brooklyn Museum, New York: p. 8 (*bottom*); Chau Chak Wing Museum, University of Sydney: pp. 71 (Nicholson Collection, purchased with funds from Sir Charles McDonald 1953, NM53.30), 72 (Nicholson Collection, donated by Liska Woodhouse 1947, NM48.1), 151 (University Art Collection, The Hon. R. P. Meagher bequest 2012, UA2012.712); photo Courtney Celley/USFWS: p. 38; from Charles Delon, *Cent tableaux de géographie pittoresque, avec une introduction topographique* (Paris, 1881): p. 176; from Charles Dickens, *Bleak House* (London, 1853), photo University of California Libraries: p. 164; Flickr: pp. 78 (Arian Zwegers, CC BY 2.0), 187 (Joey Verge, CC BY 2.0); Gallerie degli Uffizi, Florence: pp. 147, 148; iStock.com: pp. 12–13 (gustavofrazao); Kimbell Art Museum, Fort Worth, TX: p. 68; Library of Congress, Washington, DC: pp. 6, 95, 158, 169, 170; Los Angeles County Museum of Art (LACMA): pp. 8 (*top*), 116; Manchester Art Gallery: p. 69; The Metropolitan Museum of Art, New York: pp. 47, 49, 50, 52, 55, 56, 60, 62, 64, 70, 73, 145, 155 (*top* and *bottom*), 157; Musée des Beaux-Arts de Lyon: pp. 204–5; Musée d'Orsay, Paris: p. 146 (*bottom*); Museo Nacional del Prado, Madrid: p. 198; NASA's Goddard Space Flight Center: pp. 31 (Ocean Biology Processing Group), 101 (MODIS Rapid Response Team), 166 (LARC/JPL, MISR Team); National Archives at College Park, MD: p. 88; National Meteorological Library and Archive, Exeter (Open Government License, CC BY 4.0): p. 105 (F_B_8_1805–1807); New York Public Library: p. 144;

Österreichische Nationalbibliothek, Vienna (Cod. Ser. n. 2644, fol. 58r): p. 10; Palazzo Apostolico, Vatican City: p. 16; Pixabay: pp. 102–3 (ofjd125gk87), 153 (Dae Jeung Kim), 156 (morn_japan), 162 (Eric Neil Vázquez); from Gregor Reisch, *Margarita philosophica* (Freiburg im Breisgau, 1503): p. 117; Schloss Mosigkau, Dessau-Roßlau: p. 74; Shutterstock: pp. 163 (Moviestore), 171 (Amblin/Universal/Warners/ Kobal); Tokyo National Museum: p. 91; Unsplash: pp. 9 (Cassie Boca), 23 (Benjamin Sadjak), 26 (Thomas Lipke), 27 (Jaanus Jagomägi), 35 (Lison Zhao), 39 (Felix Mittermeier), 41 (Lucas Ludwig), 98 (Bruno Martins), 107 (Michael Jin), 108 (John Middelkoop), 109 (NOAA), 111 (Nikolas Noonan), 115 (Jamie Street), 122 (Lawrence Hookham), 126 (Patrick Bald), 132–3 (Zoltan Tasi), 178 (Magda V.), 184 (Jisoo Kim), 185 (Fer Nando), 199 (Tati y Adri), 200 (Mikk Tõnissoo); U.S. Marine Corps photo by Cpl. Brandon Suhr/USMC 13th MEU: p. 152; U.S. Navy photo by MC2 Jon Dasbach: p. 181; Van Gogh Museum, Amsterdam (Vincent van Gogh Foundation): p. 146 (*top*); Veneranda Biblioteca Ambrosiana, Milan: p. 21; Walters Art Museum, Baltimore, MD: pp. 77, 79, 86 (MS W.73, fol. 2r), 150; Wellcome Collection, London: pp. 57 (MS 49, fol. 6v), 192; Wikimedia Commons: pp. 20 and 24 (Famartin, CC BY-SA 3.0), 30 (Αναστασία Πορτνά, CC BY-SA 4.0), 34 (Peter Fitzgerald, CC BY-SA 4.0), 36 (Yinan Chen, public domain), 42 (Ramakrishna Gundra, CC BY-SA 4.0), 43 (Loadmaster (David R. Tribble), CC BY-SA 3.0), 44 (Dr Osama Shukir Muhammed Amin FRCP(Glasg), CC BY-SA 4.0), 67 (Sailko (Francesco Bini), CC BY 3.0 – Musée de Tessé, Le Mans), 154 (© Marie-Lan Nguyen, CC BY 2.5), 175 (NaturesFan1226, CC BY 3.0), 191 (Pierre André Leclercq, CC BY-SA 3.0), 196–7 (Leonidas (Christian Georg Becker), public domain); from John Wilkes, ed., *Encyclopaedia Londinensis; or, Universal Dictionary of Arts, Sciences, and Literature*, vol. XVI (London, 1819), photo Smithsonian Libraries, Washington, DC: p. 149.

INDEX

Page numbers in *italics* refer to illustrations